Developing Synthetic Transport Systems

Alexey Melkikh · Maria Sutormina

Developing Synthetic Transport Systems

Springer

Alexey Melkikh
Institute of Physics and Technology
Ural Federal University
Yekaterinburg
Russia

Maria Sutormina
Institute of Physics and Technology
Ural Federal University
Yekaterinburg
Russia

ISBN 978-94-007-5892-6   ISBN 978-94-007-5893-3 (eBook)
DOI 10.1007/978-94-007-5893-3
Springer Dordrecht Heidelberg New York London

Library of Congress Control Number: 2012953372

© Springer Science+Business Media Dordrecht 2013
This work is subject to copyright. All rights are reserved by the Publisher, whether the whole or part of the material is concerned, specifically the rights of translation, reprinting, reuse of illustrations, recitation, broadcasting, reproduction on microfilms or in any other physical way, and transmission or information storage and retrieval, electronic adaptation, computer software, or by similar or dissimilar methodology now known or hereafter developed. Exempted from this legal reservation are brief excerpts in connection with reviews or scholarly analysis or material supplied specifically for the purpose of being entered and executed on a computer system, for exclusive use by the purchaser of the work. Duplication of this publication or parts thereof is permitted only under the provisions of the Copyright Law of the Publisher's location, in its current version, and permission for use must always be obtained from Springer. Permissions for use may be obtained through RightsLink at the Copyright Clearance Center. Violations are liable to prosecution under the respective Copyright Law.
The use of general descriptive names, registered names, trademarks, service marks, etc. in this publication does not imply, even in the absence of a specific statement, that such names are exempt from the relevant protective laws and regulations and therefore free for general use.
While the advice and information in this book are believed to be true and accurate at the date of publication, neither the authors nor the editors nor the publisher can accept any legal responsibility for any errors or omissions that may be made. The publisher makes no warranty, express or implied, with respect to the material contained herein.

Printed on acid-free paper

Springer is part of Springer Science+Business Media (www.springer.com)

Contents

1 **Biological Cybernetics and the Optimization Problem
   of Transport of Substances in Cells** 1
   1.1 Introduction 1
   1.2 Methods of Optimization and Living Systems 4
        1.2.1 Control Theory and Biosystems 4
        1.2.2 Optimality and Living Systems 9
        1.2.3 Compartmental Models of Living Systems
               in Biological Cybernetics 10
        1.2.4 Biological Cybernetics, Synthetic Biology
               and the "Minimal Cell" 11
   1.3 Transport of Ions Through Cell Membranes:
        Models and Methods of Optimization 13
        1.3.1 Active and Passive Transport of Ions,
               Resting Potential 13
        1.3.2 Osmotic Pressure of Solutions Inside and
               Outside the Cell 16
        1.3.3 Classification of Models of Ion Transport,
               Two-Level Model, Algorithm "One Ion-One
               Transport System" 17
        1.3.4 Methods of Optimizations and Transport
               of Substances 26
        1.3.5 Two Transport Systems for One Substance 27
        1.3.6 An Optimization of the Transport System
               of a Cell as a Game Problem 31
   References 32

2 **Models of Ion Transport in Mammalian Cells** 35
   2.1 Introduction 35

2.2	Cardiac Cells		36
	2.2.1	Model of Transport Systems	38
	2.2.2	Regulation of Ion Transport	43
2.3	Neurons		46
	2.3.1	Model of Transport Systems	48
	2.3.2	Model of Ion Transport with a Restriction of Deviation from the Experimental Data	53
	2.3.3	Regulation of Ion Transport	56
2.4	Erythrocytes		59
	2.4.1	Model of Ion Transport	59
	2.4.2	Model of Regulation of Ion Transport: Efficiency or Robustness?	63
2.5	Hepatocytes		67
	2.5.1	Model for Ion Transport	67
	2.5.2	Regulation of Ion Transport	70
2.6	Regulation of Ion Transport in Compartments of a Mammalian Cell		71
	2.6.1	Mitochondria	72
	2.6.2	Sarcoplasmic and Endoplasmic Reticulum	77
	2.6.3	Synaptic Vesicles	78
2.7	Conclusions		81
References			81

3 Models of Ion Transport and Regulation in Plant Cells and Unicellular Organisms ... 85

3.1	Introduction		85
3.2	Archaea		86
3.3	Diatomei		92
3.4	*E. coli*		103
	3.4.1	A Transport Model of Basic Ions in *E. coli*	103
	3.4.2	Calculation of the Osmotic Pressure Differences in Bacteria	111
3.5	Regulation of Ion Transport in Select Microorganisms		112
3.6	Possible Regulatory Strategies for Bacterial Transport of Heavy Metals		115
3.7	Plant Cells		118
3.8	Vacuoles		122
3.9	Thylakoid		125
3.10	Conclusion		128
References			128

4	**Optimization of the Transport of Substances in Cells**	131
4.1	Optimization Methods Used for Models of Transport Subsystems of Living and Artificial Cells.	131
	4.1.1 Effectiveness of the Energy Conversion in the Transport of Substances Through Biomembranes.	132
	4.1.2 Synthesis of the Transport System of an Artificial Cell Based on the Method of Dynamic Programming. . .	137
	4.1.3 Ideal Transport System: Simultaneous Optimization of Robustness and Effectiveness	142
	4.1.4 Method of the Critical Point	145
	4.1.5 Controllability and Paradox of Ions Transport.	148
	4.1.6 Cascades and Networks of the Transport Molecular Machines.	151
	4.1.7 Regulation of the Pressure in Generalized Cells, Cells in Fresh and Distilled Water, Transport of Water	156
	4.1.8 Transport of Ions with a Lack of Energy and Diffusion of ATP.	160
4.2	Protocells at the Early Stages of Evolution	162
	4.2.1 Early Stages of Evolution and Origin of the First Cells	162
	4.2.2 A Model of the Simplest Transport System in a Minimal Cell	164
	4.2.3 A Model of the Simplest System for the Control of Transport Processes in a Cell	166
	4.2.4 Physico-Chemical Models of Cellular Movement	169
	4.2.5 Sunlight as a Possible Source of Energy for Movement.	175
	4.2.6 The Energy Balance in Protocells	177
	4.2.7 The Problem of Control and Reception of Information: Strategies Used by Protocells for Directed Motion.	178
4.3	The Transport of Large Molecules in Living and Artificial Cells.	182
4.4	Conclusion	187
References		194
Index		199

Chapter 1
Biological Cybernetics and the Optimization Problem of Transport of Substances in Cells

The methods of biological cybernetics are used to observe living systems. The models of transport of substances through the biomembrane are discussed. Compartmental models of living systems in control theory are considered. The problem of the "minimal cell" is discussed. The place and role of models of the transport of substances in synthetic and systems biology are discussed. The "one ion—a transport system" algorithm and the game approach, which were previously proposed by the authors to model the transport of ions in cells, are described.

1.1 Introduction

Living systems are optimal. This statement, of course, needs to be clarified. In what sense do we understand optimality? To what extent are they optimal? If they are not fully optimal, why not? These issues of living systems also pertain to biological cybernetics.

Biological cybernetics is the application of cybernetics to living systems. Cybernetics is traditionally composed of the following components: information theory, automata theory, control theory, operations research, pattern recognition and algorithms theory. At present, cybernetics also includes other sciences, but only the aforementioned six parts have a special mathematical apparatus that is inherent to only this science.

Biological cybernetics is the fundamental basis of systems biology. Biological cybernetics objects are living organisms, their populations, and subsystems of organisms and cells. One of the objects of biological cybernetics is the transport subsystem of the cell.

Processes involved in the transport of substances across the biomembranes of cells are important for vital cellular functions. For example, many cells (bacteria, archaebacteria, cyanobacteria, and yeast) survive considerable changes in the concentration of ions in the environment. It is necessary to understand the mechanisms by which cells resist such changes. The transport of ions is controlled, to a certain extent, in all cells. The purpose of the regulation of transport of substances is, in most cases, the maintenance of the constancy of the intracellular environment. Moreover, transport systems have one more important property: their efficiency in characterizing the ability of the cell to perform various types of useful work. However, these two properties have almost never been treated together in the context of a transport subsystem.

The goal of systems biology is a comprehensive description of a cell (or an organism) through mathematical methods by the use of computers. At present, systems biology is concerned mainly with the gene and metabolic networks of cells. However, the transport subsystem of the cell has not been studied sufficiently in systems biology.

Additionally, advancing synthetic biology requires a general approach to simulating cell processes. In particular, it is necessary to understand the general principles according to which the transport subsystem of the cell is constructed. In the future, such an understanding will help to answer why the transport subsystem of a cell is arranged exactly the way it is.

Although there are a great number of papers devoted to the application of control theory to biosystems, there has not been much discussion of the optimization of transport processes in cells. In addition to other subsystems, the transport subsystem of a cell is within the purview of systems biology [see, e.g. (Jamshidi and Palsson 2006, El-Shamad et al. 2002)]. However, optimization methods have rarely been applied to the system of the transport of substances in the cell. This is most likely because the processes of the transport of substances are assumed to be secondary to gene or metabolic processes in many respects.

The cell transport subsystem is connected with other subsystems. For example, the processes of transcription and translation are related to the transport of nucleotides and proteins through the nuclear membrane. The functioning of genetic and metabolic networks is related to the transport of proteins and metabolites within the cell and through the membranes of intracellular compartments (mitochondria, chloroplasts, nucleus and others). On the one hand, transport processes in turn depend on metabolism and other subsystems. Thus, considering the transport subsystem independently of the rest of a cell is an approximation. On the other hand, the consideration of the transport subsystem alone allows a better understanding of the other subsystems with which it interacts. Ultimately, a separate description of the transport subsystem contributes to an understanding of the laws of cellular functioning as a whole.

In addition, the construction of models for the optimization of the transport system is topical because at the early stages of evolution, the transport of substances might be one of the few functions of a protocell. An understanding of the mechanisms by which transport is optimized in elementary cells would be helpful in the creation of artificial cells.

This book is a study at the intersection of the three sciences: physics, cybernetics, and biology (Fig. 1.1).

1.1 Introduction

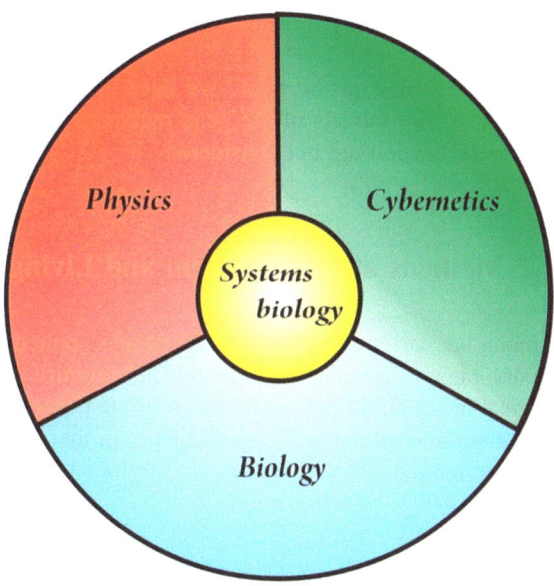

Fig. 1.1 Systems biology are the intersection of the three sciences

Biology provides information about the structure and functions of an organism (or cell); physics provides the laws, which allow us to build a model of transport processes; and cybernetics provides methods for the optimization of those processes and systems. Our goal is to present the algorithms and models of the transport subsystem of cells. These algorithms and models can maximize both the energy efficiency of transport processes and the independence of the internal environment of cells from the external environment.

The book is organized as follows. This chapter provides an overview of the methods of cybernetics as they are applied to biological systems. We discuss the place of the transport subsystem in an integrated model of the cell in terms of systemic and synthetic biology. Some of the models of ion transport in biological membranes of cells are also considered.

The Chap. 2 is devoted to models of ion transport in some mammalian cells and their compartments. These models include the first approximation (without regulation) and the regulation of ion transport at changing the composition of the environment.

The Chap. 3 is devoted to models of ion transport in the cells of protozoa, plants, bacteria, and some of the compartments of these cells. We consider the behavioral strategy of cells in response to a substantial change in environmental conditions, such as salt stress.

The Chap. 4 is devoted to models of ion transport in artificial cells. Particular attention is given to the application of optimization techniques to the transport subsystem of cells. The systems of transport of protocells in the early stages of evolution are discussed. The conditions under which various forms of energy conversion (including directed movement) would be beneficial in a protocell are discussed.

Note also that in the book, we have focused on models of the transport of substances, but models of the gene regulation associated with the transport subsystem of the cell are not considered.

1.2 Methods of Optimization and Living Systems

Among the various optimization techniques applied to biological systems, control theory plays a special role. Historically, control theory began with the mathematical modeling of the optimization of the properties of living organisms. Let us consider the basic laws of control theory in the application of biological systems, bearing in mind that they are largely common to biological cybernetics as a whole.

1.2.1 Control Theory and Biosystems

The essence of control theory is that the analyzed system is represented as a control system—the connection of individual elements in a configuration that provides the specified characteristics. The control system consists of a control device and the object of control. Control actions are aimed at achieving a certain result—the objective of control.

For biological systems, this definition means that the work of organs, tissues, and cells is conveniently represented as a separate control system and control device. Because of the large amount of feedback regulating the activity and rate of synthesis of enzymes, the concentration of their final products remains almost unchanged even when faced with a fairly wide range of external perturbations. These mechanisms are incorporated into a regulated system; thus, the term "internal control" is used [see, for example (Novoseltsev 1978)].

Internal control is considered passive. This means that the existing system, which maintains the steady state of equilibrium (or the appearance of it) as a characteristic response of the system to an external perturbation, does not require any metabolic work (Waterman 1968). Passive mechanisms of regulation are not unique to the biochemical level of organization of biosystems. For example (Novoseltsev 1978), the maintenance of the normal spatial orientation of fish provides a passive mechanism of regulation: the center of buoyancy and the center of gravity do not coincide. Thus, when a deviation of an axial plane of the fish occurs because of vertical torque, the body returns to its normal position. Passive control of the system occurs as a result of the interaction of the elements that make up "the system itself". Bertalanffy (1973) suggested the term "dynamic interaction" or "primary regulation" for this type of control.

1.2 Methods of Optimization and Living Systems

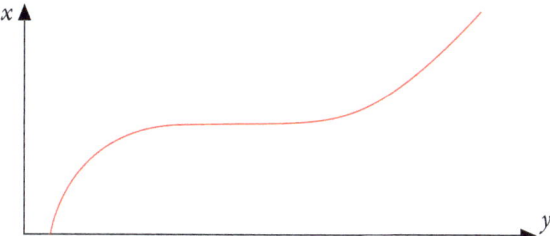

Fig. 1.2 Homeostatic curve. Here, x—is an internal parameter of an organism, and y—is an external parameter

In control theory, the decision of whether this system is a controllable system shall be made on the basis of the presence in the description of the feedback system. If feedback exists, the system is controllable. Passive systems in control theory are usually considered to be uncontrollable. In theoretical biology, however, passive mechanisms are referred to as control systems. The term "passive control system" was introduced by Ashby (1954).

For physiological systems, for example, the interaction of active and passive regulation gives rise to a characteristic dependence of the variables of the internal environment on the surrounding environment—the so-called "homeostatic curve" (Fig. 1.2).

Active control is specific only to biosystems and requires that the metabolic costs of the biosystem, which are performed by special arrangements, be separate from the elements of the system that they control. External mechanisms are designed so that any perturbation v to the system, which alters the biosystem, is detected; the detection occurs either directly or, for those effects in the system that caused this disturbance, by means of special sensors. The signal from the sensor goes to external circuit elements, where it is processed, resulting in a response that is produced by the control signal, counteracting the influence of disturbances on the system.

In Fig. 1.3, a simple external circuit of active control is applied to the internal mechanism of passive control of the metabolic structure of biological systems (Waterman 1968).

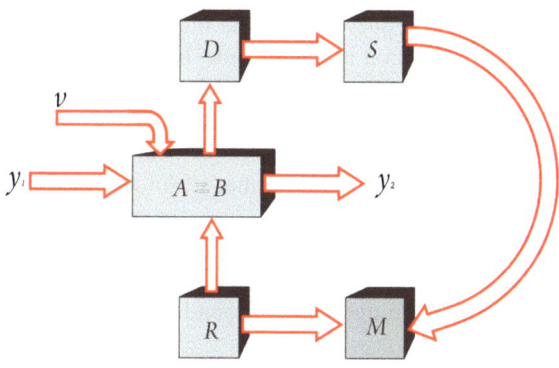

Fig. 1.3 An example of simple circuit of active control

For example, the transport of substances in the cells as an external perturbing signal is a frequently represented concentration of a substance in the environment. Receptors for those substances are located on the surface of the cell membrane (or the membranes of intracellular compartments) and signal their presence. An example of such a signal component is the calcium ion. Increasing concentrations of this ion in the cell (from the opening of calcium channels) leads to the launch of various regulatory mechanisms. Some of these mechanisms will be discussed in this book.

Sometimes, the active control processes in biological systems are realized by only one type of negative feedback—negative feedback on the *unbalance* (Bertalanffy 1973). However, this view may be only a first approximation in the study of regulation in biological systems.

In the mathematical modeling of biological systems (in particular, the description of transport processes in the cells), we have to deal with two types of variables: flow rates and levels. Levels are related, for example, to the concentrations of various substrates or enzymes in the cells. Rates characterize the speed of processes. According to Novoseltsev (1978), levels are connected with rates by an equation for the conservation of various quantities that can be written as follows:

$$\frac{dx}{dt} = \sum_{k=1}^{r-s} y_k - \sum_{k=r-s+1}^{r} y_k,$$

where x is the level of a substance in the biosystem, $y_1, y_2, \ldots, y_{r-s}$ are the rates of intake of the substances into the system or of their synthesis inside the system, and $y_{r-s+1}, \ldots, y_r$ are the rates of consumption or removal from the system.

From the viewpoint of nonequilibrium thermodynamics, levels are associated with thermodynamic forces and the rates with thermodynamic flows.

Under the purpose in a biological system is generally understood that final state in which it comes because of its structural organization, or the expected result of its operation (Anokhin 1970). Note that the question of purpose in the control of living systems is not clear. It is clear that living systems have a hierarchy of purposes, which begins with maintaining the concentration of a substance in the cell and ends in the biosphere as a whole. In the study of the efficiency of transport processes, this question is quite important and will be discussed below.

Much attention is paid in biocybernetics to the quality of functioning of living systems—their effectiveness, efficiency and reliability. The term "optimal systems" is often used. Note that the terms "efficiency" and "reliability" often require contradictory conditions for their implementation. As shown below, these terms are also "complementary" to each other to some extent, when applied to the transport of substances.

When there is a balance in the rates and flows of substances, it is possible to achieve the second order goal: the creation of those conditions of functioning within a system that do not depend on changes in the environment—*homeostasis*.

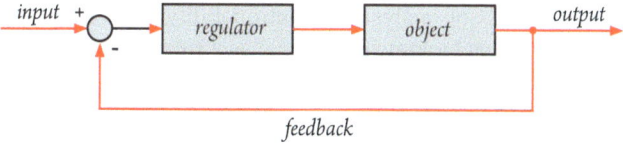

Fig. 1.4 The scheme of automatic control of the deviation proposed by Wiener, as a model of homeostatic regulation in organisms

The objectives of the third level include improving the quality of the internal environment of the system—increasing the energy efficiency, improving the design of elements, etc. (Novoseltsev 1978). However, that hierarchy of goals, which can also be applied to transport processes in cells, is not universal because it essentially depends on whether the environment is strongly variable. Obviously, in a constant environment, for example, in the depths of the ocean, the maintenance of the constancy of the internal environment of the cell is not a very urgent task. The effectiveness of processes is the primary role in this situation.

Wiener (1961) suggested that the constancy of the internal environment can be provided by a simple automatic control: negative feedback from deviations from the setpoint (desired value) (Fig. 1.4).

Subsequently, a number of other terms were introduced in addition to the term "homeostasis". For example, the term "resilience" describes the weak dependence of ecological systems on the environment (Holling 1973; Casti 1979). In accordance with the provisions of the traditional theory of dynamical systems, the term "stability" implies an ability of the system to return to its original state after a disturbance. The term "resilience" also implies the preservation of relationships within the system; it is a measure of the ability of a system to compensate for changes in state and environmental variables and thus to remain the same. This can occur, for example, by switching to a new equilibrium state. The term "inertia" was used for populations (Murdoch 1973).

One important concept in control theory is encompassed by the term "controllability". This term is applicable when the methodology changes from "input–output" to "input-state-output". This change means that for the object under study, some variables may not appear at the output. This type of approach is important for control in biological systems [see, for example, (Novoseltesv 1978)]. This raises, for example, the following question: under what conditions is it possible to transfer the system from a given initial state to the desired state in a finite time? For example, a linear control system is completely controllable if a control exists that takes a finite time to transfer the system from any initial state to a given final state. If we discuss the system of transport of substances in the cells, the controllability of such a system is a very important property because there may be situations in which the desired values for the concentrations of a given structure of the transport system cannot be achieved *in principle*. In this case, the system is not completely controllable.

The classical methods of analysis of controllability were developed for linear control systems; however, biological systems, as a rule, are not linear systems. In particular, this nonlinearity applies to the transport of substances.

One important concept, which was introduced relatively recently, is biological robustness.

Robustness is one of the fundamental characteristics of biological systems. Specific examples are given in reviews on robustness (Kitano 2004; Stelling et al. 2004; Shoemaker et al. 2010). Robustness is similar in many respects to the previously introduced term "flexibility". According to Kitano (2004), robustness is a feature that allows the system to maintain its function despite internal or external disturbances. Moreover, robustness refers more to the functions of the system than to the states, which distinguishes robustness from stability or homeostasis.

However, despite the importance of this concept in systems biology, there is no mathematical justification for this process. We must understand the fundamental principles upon which the living system is based. Creating a mathematical theory of robustness is one of the key tasks of systems biology (Kitano 2007).

The notion of "robustness" also has an important evolutionary aspect. Some species receive their robustness because of instability of the population. For example, the HIV virus is resistant to therapeutic effects because of mutations.

In control theory, there are methods of robust control. Robust control implies uncertainty in models. In this case, the control will be such that it continues to work, despite the inaccuracy in the model of the system. Thus, robust control is important with respect, for example, to the transport of substances, because an accurate model of the transport system cannot always be constructed even for the simplest of cells.

Carlson and Doyle (2002) consider another important concept that characterizes living systems—fragility. According to Csete and Doyle (2002), there is a link between robustness and fragility. HOT (highly optimized tolerance) theory states that when a system is optimized for certain perturbations, it inevitably demonstrates fragility against unexpected perturbation. As an example, this situation can be compared to the existence of species-cosmopolitans (living almost everywhere) and species-narrow specialists (optimized for a small ecological niche).

To some extent, the notion of "fragility" resembles the notion of "sensitivity" used in control theory. We can consider the sensitivity of a cell in relation to the presence of multiply charged ions as a type of "fragility" of the system of transport. This sensitivity is expressed in the fact that in the presence of multiply charged ions in the environment, the resting potential of a cell should be significantly reduced. The sensitivity of a cell to the emergence of multiply charged ions in the environment will be discussed in more detail in Chap. 4.

The term "persistence" is also applied to models of biological systems [see, for example, (Angeli and Sontag 2010)]. "The persistence property for differential equations defined on nonnegative variables is the requirement that solutions starting in the positive orthant do not approach the boundary of the orthant. For chemical reactions and population models, this translates into the non-extinction property: provided that every species is present at the start of the reaction, no species will tend to be eliminated in the course of the reaction". Thus the term "persistence" is in many aspects similar to the term "robustness".

1.2.2 Optimality and Living Systems

Optimality in biological systems is understood differently by different authors. For example, Rashevsky suggested the principle of an "optimal structure of organism" (Rashevsky 1960), which suggests that the structure can exist with the lowest consumption of metabolic energy necessary to maintain certain elements of this structure (which is necessary and sufficient for the needs of the organism).

Evolutionary optimality implies that organisms in the process of evolution reached their optimal designs. However, it is not obvious that such a process always has time to occur. Nevertheless, evolutionary optimality plays an important role in understanding other types of optimality and will be discussed below.

In (Novoseltsev 1978), it was noted that, for example, an optimal technique often provides only a 10–15 % gain compared to non-optimal designs. In addition, the closer a system is to the optimal system, the more complex it is, as a rule. Complexity, in turn, may lead to an increase in errors in the functioning of the system. For example, the problem of complexity in relation to the evolution of intracellular organelles is discussed in (Zerges 2002). The author uses examples to show that the complexity of bacteria did not allow their descendants to integrate fully into the physiology of eukaryotes (including from the point of view of transport processes).

In addition, an optimal system can only be constructed if all information about the environment is known. This information is not always available. Consequently, there must be some compromise between improving the quality of operation and energy consumption. These important issues will be considered when modeling the transport of substances in protocells.

Note that it is more realistic to find particular effectiveness criteria when modeling the most simple of biosystems—protocells (in particular, their transport properties)—than it is for complex organisms. These particular effectiveness criteria will be considered in Chaps. 2–4.

There is also a "principle of maximum simplicity", according to which a particular structure or design that we find in nature, is the simplest possible structure capable to perform this function. However, the simplicity is an indefinite term. It can be replaced, for example, by the minimum of material and energy spent.

In living nature, it is difficult to formulate a well-defined objective function. It is not always clear what constitutes an object and thus, a controller. Because all living systems are open, it is difficult to define their borders. Thus, partial quality criteria can be difficult to determine. In this regard, the most important criterion may be considered an evolutionary one: the most optimal organisms are those that replicate faster in equivalent conditions.

As noted above, a weak dependence of the parameters of the biosystem on the environment and the effectiveness of such systems are different concepts for modeling biological systems. The system can be almost independent of the external environment and at the same time be ineffective. These two concepts have not previously been considered together in relation to the transport processes in cells. As shown below, robustness and efficiency are complementary to each other: high expenses are necessary to achieve high robustness (the work of the cell at the same time will not be effective), but a system with 100 % efficiency cannot be robust. However, the concept of efficiency should be clarified. This clarification is made in Sect. 4.

1.2.3 Compartmental Models of Living Systems in Biological Cybernetics

In the description of biological systems, it is often possible to identify some relatively independent volumes involved in the transfer process as a whole. In particular, these independent volumes are often found in the transport of substances in the cell. Often, it can be assumed that some substances are uniformly distributed within the cell or in intracellular compartments. In this case, the concentration gradients and electrical potentials mainly exist on the membranes of cells or of their compartments. In the control theory of biological systems, this approach is called "compartmental analysis" (Jacquez 1972).

In general, a compartmental model contains several interconnected compartments in which two types of processes take place—the exchange of components between the individual compartments and chemical reactions that transform one component into another. The processes in compartmental models are described by ordinary differential equations. In contrast, if the system is described by partial differential equations, then the model is not compartmental. Those models are called "flow systems". For the following simulation of the transport of substances in the cells and their compartments, we will use compartmental models.

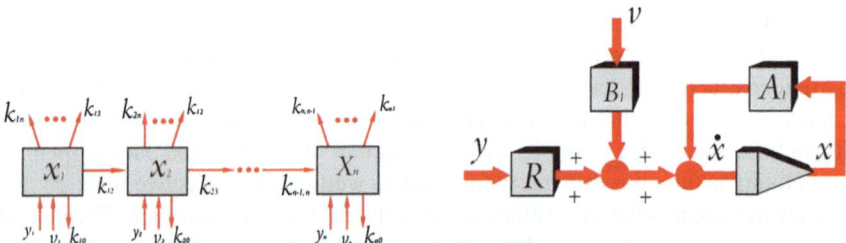

Fig. 1.5 The general scheme of a compartmental model; x_i—are the concentrations of components in compartment, y_k—are the input flows of components, k_{ij}—are the coefficients of diffusion rate, A, B, R—are matrices

The figure shows the general scheme of a compartmental model of a biological system, according to Novoseltsev (1978) (Fig. 1.5).

1.2.4 Biological Cybernetics, Synthetic Biology and the "Minimal Cell"

Currently, the application of the methods of biological cybernetics to biosystems modeling continues to evolve. For example, Senachak et al. used a slightly modified game theory to model biochemical systems (Senachak et al. 2007). The changes were related to the discrete and dynamic representation of the Nash equilibrium for games with no structural constraints and arbitrary values of gain.

The monograph by Iglesias and Ingallis examines the current state of control theory in connection with systems biology (Iglesias and Ingalls 2009). The book is mainly devoted to the analysis of the biochemical and metabolic networks of cells. The stability of linear systems is analyzed from the viewpoint of control theory using the Laplace transform. The robustness of the metabolic networks with limit cycles and the oscillations in gene networks are considered. However, ion transport is not considered at all. This area is the focus of our book.

Note, however, that (Iglesias and Ingalls 2009) only examined robustness and the few areas of control theory associated with linear theory and the analysis of stability. Optimal control theory was not considered. This is a very significant omission because the energy costs of the cell (organism) play a crucial role in its survival.

The aim of synthetic biology is to create artificial organisms or their parts. This is now a rapidly developing area; the main focus is the creation and control of artificial genes. Systems biology is used as a theoretical framework for synthetic biology.

According to Cantone et al. (2009), "Systems biology approaches are extensively used to model and reverse engineer gene regulatory networks from experimental data. Conversely, synthetic biology allows 'de novo' construction of a regulatory network to seed new functions in the cell".

The peculiar symbiotic relationship between systems and synthetic biology is discussed in the article (Smolke and Silver 2011). "Synthetic biology aims to make the engineering of biology faster and more predictable. In contrast, systems biology focuses on the interaction of myriad components and how these give rise to the dynamic and complex behavior of biological systems". The authors discuss the synergies between these two fields.

In article by Chiarabelli et al. (2009), the term "chemical synthetic biology" is discussed. According to the authors, "The term chemical synthetic biology defines that part of the field that, instead of assuming an engineering approach based on genome manipulation, is oriented towards the synthesis of chemical structures alternative to those present in nature".

In the article by Haseloff and Ajioka (2009), the place and role of synthetic biology are defined as follows: "synthetic biology is an emerging field that seeks to employ engineering principles to reprogramme living systems". This also proves that the optimization techniques, known in engineering applications should be applied to cellular subsystems. The article by Elowitz and Lim (2010), Gibson et al. (2010) reviews the achievements of modern synthetic biology and its relationship to engineering principles.

One relevant, well-known engineering principle is modularity. Many engineering systems are built according to this principle—it is often not necessary to create a new system from zero, if the restructuring of the existing lower level units is sufficient. Of course, this is not always possible. In their paper (Del Vecchio and Sontag 2009) analyzed the conditions under which the modular principle can be applied to create biomolecular networks.

Thus, an important question is whether the transport sub-system of the cell is built based on a simpler module (possibly, the transport networks of protocells). This issue will be discussed in Chap. 4.

According to Solé et al. (2007), "Cellular life cannot be described in terms of only DNA (or any other information-carrying molecule) nor as metabolism or as a compartment (cell membrane) alone. Cellular life emerges from the coupling among these three components". This observation is important because it follows that the compartmental aspect of the cells (i.e. the transport of substances) is not less important than other two.

Synthetic biology makes extensive use of both computational methods and mathematical tools. These include, for example, computational cell biology (Slepchenko et al. 2002), graph theory, Boolean logic, ordinary, partial and stochastic differential equations. In the author's opinion, "there is still no clear distinction in the scientific community between synthetic biology and somewhat older research fields, such as systems biology, biological or biomedical engineering and recombinant DNA techniques (Brent 2004). However, one might argue that the main distinctive feature of synthetic biology seems to regard the emphasis on design and testing via simulation of new living biochemical systems endowed with complex behaviour, followed by their experimental implementation".

The "minimal cell" has become an important concept in synthetic biology. To create artificial living systems, it is critical to understand the minimum conditions necessary for life. Is it possible to create such conditions in an artificial manner? Moreover, the creation of a minimal cell would help us to understand the mechanisms of the evolution of life on Earth.

Further complication of molecular complexes resulted in the emergence of protocells that were separated from the environment by a membrane. These simplest organisms could perform elementary operations with substances (energy transformation, transport across a membrane). In some studies, models of chemical energy transformation were elaborated (Murtas 2007, 2009). Those models include, for example, Ganti's chemotron (Ganti 2003), which consists of three autocatalytic subsystems, and some others (Rasmussen et al. 2004; Munteanu and

Sole 2006; Bedau 2010). Both chemical reactions and light energy are considered to be the sources of energy for protocells.

Solé et al. (2007) review the studies of the protocell. However, the transport of substances is not mentioned, although the transport of substances plays an important role in cell life. Without the transport of substances, the functioning of all subsystems of cells is impossible.

Along with the general lack of models of synthetic (minimal) cells is the lack of models for the transport of substances and, in particular, the application of optimization techniques to such systems. If we compare a subsystem to an engineering system, such as a network of roads, pipelines, resistances, etc., it becomes apparent that the problem of the synthesis of the transport system of substances in the cell is not fundamentally different from many other technical problems of this type.

The problem of the protocell will be considered in terms of the transport of substances in two ways: on the one hand, it is easy to identify the most general laws for the optimization of the transport of substances using the simplest models. On the other hand, the early stages of the evolution of life in which protocells had the simplest structure will be considered. In particular, we investigate the emergence of the directed motion of protocells.

1.3 Transport of Ions Through Cell Membranes: Models and Methods of Optimization

1.3.1 Active and Passive Transport of Ions, Resting Potential

The composition of any cell is essentially different from the composition of its environment. This difference is because the cell is in a nonequilibrium state, i.e. constantly receives energy from outside, while performing its vital functions. This energy is, in particular, spent on the transport of substances across the membrane. There are two types of transport—active and passive. The passive transport of substances occurs without the expenditure of energy and consists of the diffusion of substances from regions with higher concentration to regions of lower concentration. However, because an electrical potential difference exists across the membrane, it is more accurate to say that a substance moves from an area with a large chemical potential to a region with a lower chemical potential. The expression for the chemical potential of a substance in a solution with a concentration of particles n can be written as follows:

$$\mu = \mu_0 + kT \ln n, \qquad (1.1)$$

where μ_0 is a constant, which is not dependent on the concentration.

The expression (1.1) is valid for dilute solutions, i.e. solutions in which the concentration of ions is small. This approximation is certainly not always true for living cells. For example, there are single-celled fungi and archaebacteria that can live in nearly saturated solutions. For concentrated solutions, the Debye-Huckel theory is used, according to which:

$$\mu = \mu_0 + kT \ln n + kT \ln \gamma,$$

where γ is the activity coefficient, which depends on the concentration and properties of ions in solution [see, for example, (Freedman 2001)].

The equality of the chemical potentials of the substances on either side of the membrane is one of the conditions of equilibrium. When this equality occurs, the passive flow of ions through the membrane becomes zero. The formula for the passive flow of substances (positive ions) across the membrane is usually written as:

$$J = P\left(n^i \exp\left(\frac{Z\varphi e}{kT}\right) - n^o\right),$$

where n^i and n^o are the concentrations of the ions inside and outside of the cell, respectively, T is the temperature, Ze is the charge of ions, φ is the resting potential on the membrane, and P is the permeability of the membrane to the specific type of ion.

For small differences in the chemical potentials of the ions between both sides of the membrane, a linear relationship occurs between the flux J and the force X:

$$J = LX,$$

which is characteristic for linear non-equilibrium thermodynamics. However, for most situations, this relationship does not hold because the thermodynamic force for the biomembrane is great.

The active transport of ions requires energy. This energy can be either the free energy of ATP hydrolysis or the free energy of other ions. In terms of the thermodynamics of irreversible processes, active transport is a cross-section effect in which the flow is not proportional (in the linear case) to its own thermodynamic force but to another force that exists in the system. In this case, the force is the difference between the chemical potentials of the reactions of ATP $\to$ ADP + P, which can be represented as:

$$\Delta\mu_A = kT \ln \frac{n_A \, n_{D0} \, n_{P0}}{n_D \, n_P \, n_{A0}},$$

where n_A, n_D, and n_P are the concentrations of ATP, ADP and P, respectively, and n_{A0}, n_{B0}, and n_{P0} are the concentrations of these substances at equilibrium.

It is obvious that in the steady state, the flux of each type of substance must be zero (if the substance is not involved in chemical reactions in which it is consumed or produced):

$$J = \sum_i J_i = 0.$$

1.3 Transport of Ions Through Cell Membranes

The electric potential across the membrane of cells varies from approximately -10 mV for the membrane of red blood cells to approximately -240 mV for the membrane of some fungi. These potentials indicate that the internal environment of the cell has a negative potential relative to the potential of the environment, which is assumed to be zero. However, this does not mean that the electric charge is distributed uniformly within the cell. Most of the intracellular solution (and most of the solution of the environment) is electrically neutral. The uncompensated (excess) charge is concentrated in a narrow layer near the membrane. The thickness of this layer, called the Debye layer is approximately 3 Å. Thus the cell can be regarded as a charged capacitor in which the charge and potential difference are related by:

$$q = C\varphi,$$

where C is the capacity of membrane.

If the ion is transported only passively, then, sooner or later, the chemical potentials of the ions on both sides of the membrane will be equal. Thus, the concentrations of passively transported ions on both sides of the membrane are related by:

$$n^i = n^o \exp\left(-\frac{\varphi Z e}{kT}\right),$$

which is called the Nernst equation. This formula is just a special case of the Boltzmann distribution for particles in an external field.

What is the mechanism for the electrical potential difference across the membrane? Clearly, the potential difference is associated with the different compositions of the solutions on both sides of the membrane. If these solutions are the same, the difference in the electrical potential is zero. What, then, generates the difference in the compositions of the solutions on either side of the membrane? The first cause of the inequality of the concentrations (more precisely—the chemical potentials) of the ions on both sides of the membrane is the active transport of ions. For example, Na^+–K^+–ATPase in mammalian cells is the major transporter of sodium ions from the cell. When using this transport system (an ion pump), three sodium ions are removed from the cell in exchange for two potassium ions that are transported into the cell. In this case, the sodium ions move against the chemical potential gradient.

The second reason for the dissimilarity in the intracellular and extracellular composition is that there are ions inside the cell that are not found in the environment. These are so-called "non-penetrating" ions. As a rule, non-penetrating ions are synthesized within the cell based on the nutrients it receives from the outside. For example, metabolites that are synthesized by bacteria are non-penetrating ions. Despite the absence of a specific channel for these ions, there is limited permeability of the membrane for these ions; however, for calculations, it is often convenient to assume that they do not penetrate through the membrane at all. Because of the presence of these ions inside the cell, even in the case of cell death, the membrane potential is not equal to zero (when non-penetrating ions have not yet diffused into the environment). This residual potential is the so-called Donnan potential. The formula for the Donnan potential for the case when singly charged positive and negative ions in the environment passively penetrate the membrane is as follows:

$$\varphi_D = kT \ln \left(\sqrt{\left(\frac{z_a n_a}{2n_-^o}\right)^2 + 1} - \frac{z_a n_a}{2n_-^o} \right), \quad (1.2)$$

where Z_a and n_a are the charge and concentration of the non-penetrating ions inside the cell, respectively, and $n_-^o = n_+^o$ is the concentration of the negative passive ions in the environment. As a rule, this potential is not large and is equal to approximately -10 mV.

1.3.2 Osmotic Pressure of Solutions Inside and Outside the Cell

Osmotic pressure plays an important role in the life of cells. If it is too high, the cell membrane can break down. However, even if a membrane breakdown does not occur, the shape of cells can change when the internal pressure changes. For many organs and tissues (especially animals), the shape is fundamentally important. For example, the shape of red blood cells is critical to their motion in small capillaries.

For very dilute solutions, the osmotic pressure formula is very similar to the formula for an ideal gas:

$$p = kT \sum_i n_i, \quad (1.3)$$

where n_i is the concentration of the i-th type of ions.

When the electrostatic interaction between the ions is taken into account, (1.3) takes the following form:

$$P = nkT - \frac{e^3}{3(\varepsilon\varepsilon_0)^{3/2} 8\pi\sqrt{kT}} \left(\sum_i z_i^2 n_i\right)^{3/2}, \quad (1.4)$$

where ε is the dielectric constant of the environment and e is the charge of an electron. For the derivation of this formula, it is assumed that the second term is much smaller than the first. For example, using a typical solution for the cell and environment with a concentration of 100 mM for the negative and positive ions, it can be shown that the osmotic pressure is approximately 5 atm, and the amendment to it is approximately 12 %.

The solution can be considered dilute if the potential energy of interaction of the ions in this solution is much less than the thermal energy kT:

$$n \ll \left(\frac{kT4\pi\varepsilon\varepsilon_0}{z^2 e^2}\right)^3.$$

For singly charged ions at 300 K, this condition can be written as follows:

$$n \ll 5M.$$

In many cases, the total concentration of ions in the cells is approximately 0.2–0.3 M. This means that most of the solutions in cells and their environments can be considered dilute, and formula (1.4) applies to them. However, there are exceptions—archaea, cyanobacteria and some fungi live in an environment with a concentration of ions of a few M, i.e. close to saturation. In this case, formula (1.4.) cannot be true. In accordance with Freedman (2001), we shall write for the pressure

$$p = kT \sum_i n_i \gamma_i, \qquad (1.5)$$

where

$$\lg \gamma_i = -0.509 z_i^2 \left(\frac{1}{2} \sum_i z_i^2 m_i\right)^{1/2}$$

with m_i being the number of moles in 1000 g of the solvent. In this case, the amendment to the osmotic pressure associated with the activity coefficient γ of the solution can be large.

Thus, when considering models of single-celled living organisms in concentrated solutions (see Chap. 3), we will use formula (1.5).

1.3.3 Classification of Models of Ion Transport, Two-Level Model, Algorithm "One Ion-One Transport System"

In the paper by Melkikh and Seleznev (2006a), a classification of the models of ion transport was formulated. According to (Melkikh and Seleznev 2006a), the available papers dealing with active transport are reduced to three issues: the determination of the potential (Hodgkin and Katz 1949; Goldman 1943; Hodgkin and Horowicz 1959; Sperelakis 2001) using the known concentrations of ions in a cell; the discussion of the possible mechanisms responsible for nonzero values of the cross coefficients L_{iA} (Kedem and Katchalsky 1958; Caplan and Essig 1983; Kjelstrup et al. 2005); or the deduction of semiempirical formulas that describe ion flows through the membrane (which were measured experimentally) as a function of the membrane conditions (Sagar and Rakowski 1994; Kabakov

1994, 1998; Fahraeus et al. 2002; Faber and Rudy 2000; De Weer et al. 2001; Tsong and Chang 2003; Hopfer 2002).

The resting potential of cells is determined in studies of the first type, but only when there is experimental data on the concentrations of ions in cells. Membrane potentials, which are determined by this method, disregard the chemical potential difference of the ATP $\to$ ADP + P reaction, which is not zero in vivo. If the active transport does not generate the electric current, the sum $\sum_i z_i J_i$ is independent of $\Delta\mu_A$, and we arrive at the known Hodgkin-Katz formula

$$\varphi = -\frac{kT}{e}\ln\left(\frac{P_K[K^+]_i + P_{Na}[Na^+]_i + P_{Cl}[Cl^-]_o}{P_K[K^+]_o + P_{Na}[Na^+]_o + P_{Cl}[Cl^-]_i}\right),$$

where P_{Na}, P_K, and P_{Cl} are the permeabilities of the membrane to sodium, potassium and chlorine, respectively.

Regardless, the drop of ion concentrations across the membrane cannot be determined in theoretical terms with this method.

Studies of the second type do not deal with the determination of the resting potential or the concentrations of ions in a cell at all. Instead, they look for and discuss possible mechanisms for the active transport of ions. The relationship between an event of the ATP $\to$ ADP + P chemical reaction and the transport of ions through the membrane is sought, including how these mechanisms correlate with the law of degradation of energy.

Most studies of the third type are based on measurements of electric current, ion flow, resting potential, ion concentrations, and ATP. Those experiments revealed some interesting dependences. However, studies of this type generally disregard the full extent of thermodynamic limitations; therefore, one should not overestimate the capacity of this approach.

In the paper by Melkikh and Seleznev (2006a), the basic requirement for models of the transport of ions are considered. The fulfillment of these requirements will provide the most comprehensive answer to the question about the nature and values of the resting potential of a cell and the difference in the concentrations of basic ions:

1. The resting potential of cells φ, should depend on the difference in chemical potentials of the ATP $\to$ ADP + P reaction. If $\Delta\mu_A \to 0$, the resting potential φ should approach the Donnan potential observed in dead cells. Using the concentrations of ions in the environment, the model should be able to determine the concentrations of ions in a cell (currently, internal ion concentrations must be measured experimentally in vivo).
2. If the mechanism of active transport is turned off $(\Delta\mu_A \to 0)$, the Hodgkin-Katz potential, which is known from the literature, should be established at the membrane. This potential arises when the passive electric current of all of the ions in the system goes to zero. The same (Hodgkin-Katz) potential should also occur when $\Delta\mu_A \neq 0$ if the ionic pump does not produce an electric current (i.e. the electric charge is fully compensated).

1.3 Transport of Ions Through Cell Membranes

3. An analysis of biological evolution shows that the competition for survival is won by those mechanisms (and the corresponding organisms) with the largest ATP energy utilization factor. Direct experiments concerned with the study of the active transport of ions also give η values at a level of 90 % and larger (see, e.g. Bustamante et al. 2001). Obviously, theoretical models that do not satisfy large η values will fail to describe these experiments.
4. Some experiments demonstrated that typical ionic pumps have the property of reversibility, which corresponds to ATP synthesis if $\Delta\mu_A = 0$ and an actively carried ion has a chemical potential difference. The appropriate theoretical models should have the same property.
5. The model should not be thermodynamic but statistical in character, which implies a rough but still detailed description of the structure of the ATPase responsible for the active transport of ions. The model should provide values of the kinetic coefficients with small thermodynamic forces.

Previous studies, which were performed using linear or nonlinear non-equilibrium thermodynamics, have satisfied items 1, 2 and 4. However, they could not determine the potential and the internal concentrations of ions. These approaches fail to provide quantitative results because of the lack of information about the kinetic coefficients. Moreover, they do not account different types of ions and channels simultaneously, which does occur in some types of cells.

It was previously mentioned that studies of the second type are not concerned with the determination of common characteristics of cells at all.

Studies of the third type generally disregard, to some extent, the limitations listed under items 1, 2, 3, 4 and 5.

As discussed in Melkikh and Seleznev (2006a), many models of active ion transport do not meet the requirements formulated.

Models of ion transport that do meet the requirements were previously proposed by the authors Melkikh and Seleznev (2005, 2006a, b). Let us consider these models in detail.

The main provisions of the models are reduced to the following (Fig. 1.6):

- A system of the active transport of ions includes a sorption center for several ions.
- A macromolecule has two conformational states corresponding to different positions of the sorption center: one on each side of the biomembrane (inside the cell and in the environment).
- The ion-carrying macromolecule changes from one state to the other at the expense of the free energy of the ATP hydrolysis or the free energy of other ions.
- The probability of this transition is determined by the difference in chemical potentials of the ATP-ADP system or other ions.
- The transport of ions against the electrical field reduces the probability of this transition because it requires energy.

Fig. 1.6 Sketch of ion transport by an ATPase

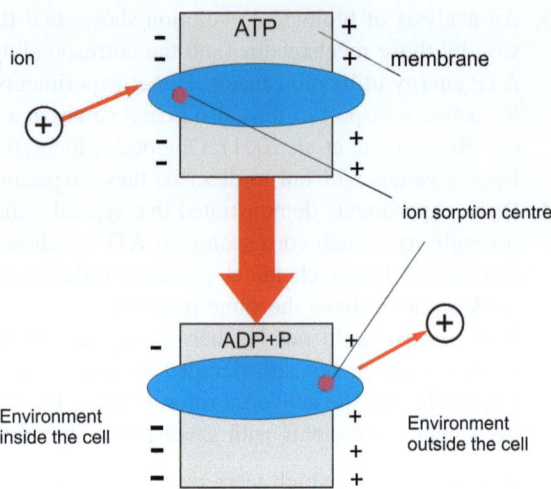

Using this model, it is possible to independently calculate both the electrical potential at the biomembrane and the concentrations of ions in a cell.

The transport macromolecule is modeled as a two-level system with a difference of Q between energy levels. In equilibrium (according to the Boltzmann distribution), the probabilities of finding the system in the upper and lower levels, respectively, can be written as follows:

$$f_k = \frac{\exp\left(-\frac{Q}{kT}\right)}{\exp\left(-\frac{Q}{kT}\right)+1}; \quad 1-f_k = \frac{1}{\exp\left(-\frac{Q}{kT}\right)+1}.$$

After an ion (ions) comes into its (their) sorption center and the presence of ATP, the free energy is transmitted to the two-level molecule. The resulting probabilities of finding the system in the upper and lower levels are equal to:

$$f_{in} = \frac{\exp\left(\frac{\Delta\mu_A-Q}{kT}\right)}{1+\exp\left(\frac{\Delta\mu_A-Q}{kT}\right)}, \quad 1-f_{in} = \frac{1}{1+\exp\left(\frac{\Delta\mu_A-Q}{kT}\right)},$$

where $\Delta\mu_A$ is the difference in chemical potentials of the ATP-ADP system.

The model suggested that when the two-level molecule is in the lower level, its position is on the inner side of the membrane, and when it is in the upper level, its position is on the outside. In this case, after the arrival of ATP and the ion in their sorption centers, the part of the macromolecule with the ion turns from the inside to the outside; the ion then goes into solution. Based on these assumptions, we can write the frequency of transfer of ions from the inside of the cell and back in the form:

1.3 Transport of Ions Through Cell Membranes

$$v_{iL \to R} = v_{iL} f_{in}(1 - f_k) p_{L \to R},$$
$$v_{iR \to L} = v_{iR} f_k (1 - f_{in}),$$

where $v_{iL \to R}$ and $v_{iR \to L}$ are the frequencies of an ion coming into the sorption center from the left (outside the cell) and from the right (inside the cell), respectively, and $p_{L \to R}$ is the probability of overcoming the electrostatic barrier for an ion.

Because the temperature inside the cell, in most cases, is equal, here and below we use dimensionless units:

$$\Delta \mu_A \equiv \frac{\Delta \mu_A}{kT}, \ \Delta \mu_i \equiv \frac{\Delta \mu_i}{kT}, \ e\varphi \equiv \frac{e\varphi}{kT}.$$

Finally, we can write a formula for the ion flux:

$$J = C\left(n^i \exp(e\varphi) \exp(\Delta \mu_A) - n^o\right). \tag{1.6}$$

where n^i and n^o are the concentrations of ions inside and outside the cell, respectively, and C is the quantity that characterizes the working speed of the transport system. The factor $\exp(e\varphi)$ reflects the fact that the transfer of positive ions from inside the cell outward requires work against the forces of the electric field.

When one ion is transferred for another (the exchanger), the structure of the formula remains the same. For example, when one positive ion is exchanged for another, the following can be written:

$$J_{1-2} = C_{1-2}(n_1^i n_2^o - n_1^o n_2^i),$$

where 1 and 2 represent two different positive ions, one of which is transferred from the cell in exchange for the other. Because the net charge is not changed, the factor $\exp(e\varphi)$ is not required.

In the steady state, all ion flows must be equal to zero. Thus, we can receive the dependencies of the concentrations of the internal ions on the external ones. For example, from (1.6), we have:

$$n^i = n^o \exp(-\Delta \mu_A + e\varphi). \tag{1.7}$$

In order for the system of equations to be closed, we also require an equation for the potential. This equation is the condition of electro-neutrality.

For example, if a cell contains a single species of positive ion and two negative ones (one of which is non-penetrating with the concentration n_a and charge z_a), we have:

$$n^i = n_-^i + z_a n_a, \tag{1.8}$$

where n_-^i is the concentration of negative ions inside a cell.

If the negative ions are carried by passive transport only, the concentrations of these ions inside and outside of the cell will be described by the Nernst relation

$$n^i_- = n^o_- \exp(e\varphi). \quad (1.9)$$

If (1.7) and (1.9) are substituted into the expression (1.8) for the stationary condition ($\Delta\mu_A$ = const), one may easily obtain the following expression for the resting potential:

$$\varphi = \ln\left(\sqrt{\left(\frac{z_a n_a}{2n^o_-}\right)^2 + \exp(-\Delta\mu_A)} - \frac{z_a n_a}{2n^o_-}\right). \quad (1.10)$$

Substituting φ, which was calculated from (1.10), into (1.7), we find the concentration of positive ions inside a cell:

$$n_{iL} = n_R^- \exp(e\varphi) + z_{an}n_{an} = n_R^- \sqrt{\left(\frac{z_{an}n_{an}}{2n_R^-}\right)^2 + \exp(-\Delta\mu_A)} + \frac{z_{an}n_{an}}{2}. \quad (1.11)$$

If $\Delta\mu_A = 0$ in (1.10), we obtain the known Donnan formula for the potential of a dead cell.

Thus, the proposed model actually fulfills the task of determining the electrical potential of the cell and the concentrations of ions inside the cell under in vivo conditions.

Note that the equations for the flows of ions can be written in an equivalent form using the language of chemical kinetics. For example, for the antiporter, which exchanges one positive ion S for another M, according to Melkikh and Seleznev (2012), we can write the following equation:

$$J_M = J_s = n_F^f k_\uparrow^o n_s^o n_M^o \frac{k_\downarrow^i}{k_\downarrow^i + k_\downarrow^o} - n_F^f k_\uparrow^i n_s^i n_M^i \frac{k_\downarrow^o}{k_\downarrow^i + k_\downarrow^o}$$

$$= \frac{n_F^f k_\uparrow^i n_s^i n_M^i k_\downarrow^o}{k_\downarrow^i + k_\downarrow^o}[\exp(\Delta\mu_s + \Delta\mu_M) - 1],$$

where n_s^i and n_M^i are the concentrations of substances S and M inside the compartment (cell), n_s^o and n_M^o are the concentrations of these substances outside, n_F^f is the concentration of the enzyme that transports the substances, and the k's are the rate constants of forward and back reactions. An equation that accounts for the dependence of transport on ATP and the electrical potential can be found similarly.

The proposed model has the property of reversibility. This means that, while maintaining a stationary concentration, the difference of ions on both sides of the membrane to ATP will be synthesized (or other ions are transported). As noted above, this property does occur for transport systems under certain conditions for many cells and their compartments.

Fig. 1.7 Two conjugated processes

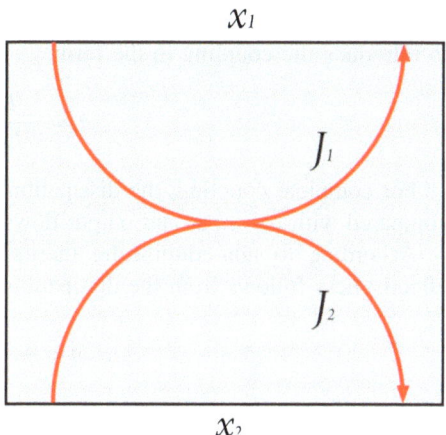

The book (Caplan and Essig 1983) discusses the efficiency of ion transport in biological membranes in terms of thermodynamics. If there are several thermodynamic forces, in many cases, it is convenient to use a dissipative function, which is the product of conjugate flows and forces. The dissipation function represents the rate of dissipation of free energy in the system.

However, a disadvantage of this approach is that the kinetic coefficients that connect the flows and forces cannot be determined within the framework of thermodynamics.

Caplan and Essig (1983) considered a simple system shown in Fig. 1.7.

Conjugation, a mechanism that may not be known in general, takes place inside a "black box". This black box usually contains the work element of energy conversion, such as a membrane in the process of active transport. Because there are two processes, the dissipation function contains two terms. In general, we can choose one process that always leads to a positive term, and it can therefore be considered as an energy source or input. The second process usually leads to a negative term and represents output. Then, the dissipation function takes the form:

$$\Phi = J_1 X_1 + J_2 X_2,$$

where the first term is the output energy and the second is the input energy. The input flux always proceeds in the spontaneous direction, that is, its direction is in accordance with the force paired to it, X_2. The output flux can also go against the corresponding force X_1. In these cases, the output term has a negative sign, indicating that the free energy is transferred from the working element to the environment. Thus, part of the free energy expended by process 2 is converted into a form characteristic of process 1. Because the dissipation function is always negative, the output flow can never exceed the input flow. It is intuitively clear that an effective energy conversion requires a high degree of coupling between these processes.

From the viewpoint of linear non-equilibrium thermodynamics, it is convenient to introduce the coupling in the form:

$$q = \frac{L_{12}}{\sqrt{L_{11}L_{22}}}.$$

For complete coupling, the dissipation function can be made arbitrarily small compared with the input and output flows of energy.

According to non-equilibrium thermodynamics, a simple definition for the effectiveness follows from the dissipation function:

$$\eta = -\frac{J_1 X_1}{J_2 X_2}.$$

Thus, the effectiveness is the ratio of the output power to the input power. When process 2 is spontaneous and process 1 is not, then

$$0 < \eta < 1.$$

If X_2, which is the input or driving force, remains constant and restrictions are not imposed on X_1, then the flow of J_1 will continue as long as X_1 reaches a value sufficient to terminate it (the state of static head). In not fully conjugated systems, energy must be expended to maintain the state of static head, even if the output flux is zero. Stationary states with $X_1 = 0$ are called states with level flow. In this case, the energy is spent, even if the output force is zero.

However, this energy expenditure is still helpful. Therefore, the effectiveness function cannot be regarded as a universal criterion. According to the authors (Caplan and Essig 1983), a universal criterion does not exist.

A common drawback of these approaches is that an effectiveness based only on thermodynamics is impossible to calculate. Only statistical models (in particular, the one proposed by the authors) make it possible to calculate the concentration of ions and the potential inside the cell independently.

In accordance with the classical definition in (Melkikh and Seleznev 2006), a formula was proposed for the efficiency of the active transport of ions as the ratio of output capacity to input capacity:

$$\eta = \frac{J' \cdot \Delta\mu^e}{v' \cdot \Delta\mu_A}, \tag{1.12}$$

where $\Delta\mu^e = \mu^{eo} - \mu^{ei}$ is the difference in the electrochemical potentials of the ions in the solutions on both sides of the membrane, which occurs during a steady-state operation of the pump. However, rather than considering this relationship with the existing steady-state flows and forces, flows with vanishing differences in the chemical potentials of the ions (v' and J') were considered. This assumption was adopted because of the fact that when the steady-state flux of ions of each type is zero, the meaning of (1.12) is unclear.

1.3 Transport of Ions Through Cell Membranes

As was shown in (Melkikh and Seleznev 2005, 2006a, b), the efficiency of a transport system in the framework of the model is equal to unity. This result is because, in the framework of the model, energy loss does not occur during the conformational change (of the two-level macromolecule) and thus during the transfer of an ion to the other side of the membrane. Indeed, if there is only one system of transport for one ion, then using equation (1.7), we have:

$$\Delta\mu_A^e = \Delta\mu_A$$

Because the transfer of an ion is rigidly connected with the synthesis of ATP, the ion flux will be equal to the ATP flux. In this case, we find that $\eta = 1$.

Of course, in real systems, the efficiency of transport cannot be equal to unity. On the other hand, as shown by Melkikh and Seleznev (2008), Melkikh and Sutormina (2011), if there are two or more transport systems that carry a single ion at the same time, their overall efficiency is always less than unity, even if each of them separately gives an efficiency equal to unity. This conclusion is important for evaluating the effectiveness of the transport system of the cell as a whole.

The reduced efficiency in the simultaneous operation of two or more systems of ion transport is not necessarily bad for the cell. As shown in (Melkikh and Sutormina 2011), the regulation of ion transport can be provided by the simultaneous operation of two or more transport systems or by switching from one transport system to another. Because maintaining constant internal ion concentrations is an important property of cells, it is quite possible that the decrease in efficiency is, nonetheless, effective.

Because most of the ions in cells and their compartments are transferred by more than one system of transport, there is a problem in solving the system of equations for the flow of ions. In this case, to obtain an expression for the internal concentration of each ion, it is necessary to know the constants characterizing the rate of each transport system. These values are not always known; in addition it would be convenient to have analytic expressions for the internal ion concentrations for a qualitative analysis of ion transport.

To address this situation, the "one ion—one transport system" algorithm was proposed in (Melkikh and Sutormina 2008). The essence of this algorithm is that for each ion, a major transport system is selected, while the rest of the systems for a given ion are considered to be regulatory. By equating all of the flows generated by the main transport systems to zero and using the electro-neutrality condition, the dependences of the internal ion concentration and the potential on the environmental ion concentrations can be obtained. This algorithm allows the analytical calculation of the resting potential and internal concentrations without requiring data on the rate of each transport system. This calculation is possible because when there is just one transport system, its rate of work in the steady state is irrelevant. This can be observed from the structure equations (1.7).

Next, we will consider the models of cells and compartments that are based on the "one ion—one transport system" algorithm and regulation.

1.3.4 Methods of Optimizations and Transport of Substances

The process of ion transport, as well as other biological processes, is optimal in varying degrees. However, the methods of biological cybernetics were essentially not applied to the transport of substances across biomembranes. One reason, apparently, is that many of the transport models contain a large number of adjustable parameters. The physical meaning of many of the parameters is often unclear, which does not allow the application of optimization methods to those models.

Another disadvantage of transport models is that most of the models cannot incorporate the dependence of internal ion concentrations and the resting potential on the environmental ion concentrations. However, it is crucial to use optimization techniques when those dependences exist; otherwise, the model will not be closed. The application of optimization techniques may be substantially limited for those systems.

In the paper (El-Shamad et al. 2002), the authors examined the regulation of calcium ion in mammalian cells. The desired concentrations of Ca^{2+} were maintained with a model of integral control with feedback as a functional module. The authors justify the choice of integral feedback to describe calcium regulation by the requirement to obtain a zero steady-state error. Calcium is tightly regulated in mammals because of the importance of the concentration of these ions to many physiological functions. In their work, a model for calcium homeostasis was developed; the model used an integral feedback control as a functional module to support calcium homeostasis. The authors proposed a simple dynamic model for the regulation of the internal calcium ion concentration, which produces results in good agreement with experimental values.

In the paper by Melkikh and Sutormina (2011), the optimization of ion transport was considered a game problem. Let us consider this approach in detail.

In the context of the transport models proposed earlier, let us consider an abstract cell with several systems for the transport of basic substances (see Sect. 1.3.3). The main problem is how to simultaneously meet the following conditions: the internal medium of the cell should be insensitive to changes in the composition of the environment, and transport processes should be sufficiently effective.

First, we shall consider the transport of one substance in the absence of a charge. This situation can take place either with neutral substances or when the concentration of these ions is small and has negligible influence on the resting potential.

As observed from (1.7), the dependence of the internal concentration on the external ones is linear in this case. Then, if only one transport system is available for each ion, a constant internal concentration for the ion cannot be ensured when the external concentration for that ion varies. However, as was shown earlier, the efficiency of such transport systems can be close to 100 %.

This assumption of a unique transport system for each ion (a hierarchical "one ion—one transport system" algorithm) is useful in many cases for incorporating the resting potential and intracellular ion concentrations (see Chaps. 2 and 3). However, it cannot be used for modeling the regulation of ion transport.

1.3.5 Two Transport Systems for One Substance

In accordance with Melkikh and Sutormina (2011), we now consider the case when two different transport systems are available for one substance. We shall show that the system can be rendered mostly insensitive to changes in the environmental concentrations by varying the capacity of these transport systems. The flux of a substance generated by the two systems can be written in the form

$$C_1\left(n^i \exp(\Delta\mu_A) - n^o\right) + C_2\left(n^i \exp(\Delta\mu_B) - n^o\right) = 0, \quad (1.13)$$

where $\Delta\mu_B$ is the motive force of the second transport system (dimensionless).

Experiments with different cells demonstrate that, as a rule, in normal conditions, each ion has one basic transport system. This transport system creates a difference between the intra- and extra-cellular concentrations of this ion that corresponds to the normal value. The other transport systems are regulatory, which means that they should operate when the environmental composition begins changing. Let us plot the relationships between the internal and external concentrations for each system to demonstrate the result of the concurrent operation of these systems. For example, let the relationships have the shapes shown in Fig. 1.8, where the circle denotes the concentration of an ion in the normal state.

It can be shown that the total efficiency of two or more different transport systems working concurrently will always be less than 100 %, even if the efficiency of each individual system is 100 %. Let us determine the efficiency when two transport systems work concurrently for one ion. Let one ion be transferred by two pumps with different values for the motive force. Then, the internal concentration of this ion can be calculated using Eq. (1.13):

$$n^i = \frac{n^o(C_1 + C_2)}{C_1 \exp(\Delta\mu_A) + C_2 \exp(\Delta\mu_B)}.$$

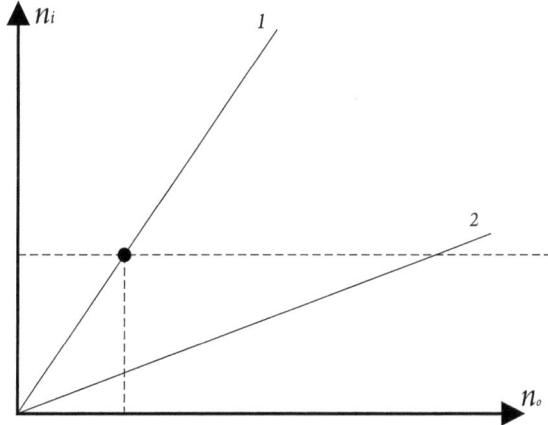

Fig. 1.8 Dependencies of internal concentrations on external concentrations

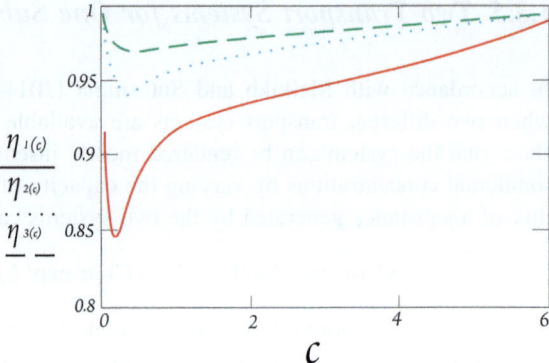

Fig. 1.9 Dependencies of the efficiency of two simultaneously working transport systems on the ratio of their work speeds

If the motive forces ($\Delta\mu_A$ and $\Delta\mu_B$) are constant, the dependence $n^i(n^o)$ is linear.

In this case (the absence of a charge), the difference in the chemical potentials of an ion can be defined as

$$\Delta\mu = \ln\frac{n^o}{n^i} = \frac{C_1 \exp(\Delta\mu_A) + C_2 \exp(\Delta\mu_B)}{C_1 + C_2}.$$

Therefore, for the efficiency, we have

$$\eta = \frac{J' \cdot \Delta\mu}{v' \cdot \Delta\mu_A}$$

$$= \frac{(C_1(\exp(\Delta\mu_A) - 1) + C_2(\exp(\Delta\mu_B) - 1)) \cdot \ln\frac{C_1 \exp(\Delta\mu_A) + C_2 \exp(\Delta\mu_B)}{C_1 + C_2}}{C_1(\exp(\Delta\mu_A) - 1) \cdot \Delta\mu_A + C_2(\exp(\Delta\mu_B) - 1) \cdot \Delta\mu_B}$$

It is easy to see that if C_1 or C_2 is zero (only one transport system operates), the efficiency is 1 at any force. In contrast, if the forces are equal, the efficiency is 1 at any C_1 and C_2. For all of the other cases, the efficiency is <100 %.

Figure 1.9 presents the dependencies of the efficiency on the ratio $C = C_2/C_1$ at different forces (from the top down: $\Delta\mu_A = 2$, $\Delta\mu_B = 2.5$, $\Delta\mu_B = 2.9$, and $\Delta\mu_B = 3.9$).

When $C \to 0$ and $C \to \infty$, the efficiency approaches unity.

The formula retains its structure irrespectively of the number of fluxes and forces:

$$\eta_i = \frac{\sum_{i=1}^{n} C_i(\exp(\Delta\mu_{Ai}) - 1) \ln \frac{\sum_{i=1}^{n} C_i \exp(\Delta\mu_{Ai})}{\sum_{i=1}^{n} C_i}}{\sum_{i=1}^{n} C_i(\exp(\Delta\mu_{Ai}) - 1)\Delta\mu_{Ai}}. \quad (1.14)$$

1.3 Transport of Ions Through Cell Membranes

In this case, the first multiplier in the numerator represents the sum of the pure active fluxes of the given ion that are produced by all of the forces $\Delta\mu_{Ai}$. The second multiplier is the chemical potential of the ion, which occurs as a result of the active transport of this ion by different systems. The denominator is the total energy expense for all of the active transport systems carrying the given ion. It can be shown that $\eta < 1$ for all conditions of (1.14).

A more complicated discussion of effectiveness in the transport of substances will be presented in Chap. 4.

The deduced formulas are easily generalized to the presence of a charge by introducing the chemical potentials of ions as basic variables.

Then, two strategies, which correspond to the limiting cases of pump operation, can be distinguished in the regulation of the ion transport.

According to (Melkikh and Sutormina 2011), in the first case, we shall require that the efficiency of two concurrently operating systems is always 100 % (an absolutely efficient system). This efficiency is possible if only one of the transport systems for one ion is working at any given moment. Then, it is easy to see that a relative robustness (a weak dependence of the internal environment of the cell on the external one) can be achieved by following the curve that most closely approaches the normal concentration (the heavy line in Fig. 1.10). Thus, the cell can maintain the internal ion concentrations required for normal life within an acceptable deviation δ from the norm $[\tilde{n}_{in} \pm \delta]$, where $\tilde{n}_{in}$ is a value equal to the normal concentration of an ion for a cell in a physiologically normal state. Therewith, the robustness will knowingly be larger than the robustness ensured with one transport system only.

Notice that this piecewise dependence of the internal concentrations on the external ones fully corresponds to the term "robustness" discussed above. This switching of transport systems is known to occur in many cells (see, for example Chaps. 2–3). Switching can occur because of the considerably nonlinear dependence of the pump operation frequency on the concentrations $C = C(n^i, n^o)$.

In the second case, we shall require that the internal concentrations are fully independent of the external concentrations, at least in some regions. This requirement is fulfilled by selecting an appropriate concentration dependence for

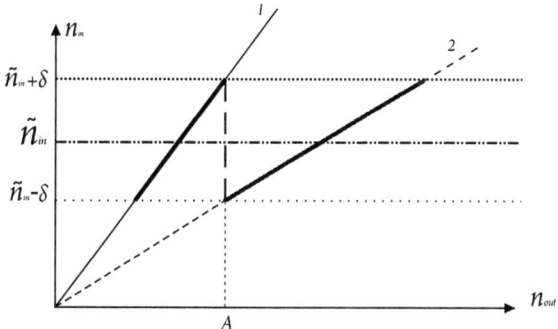

Fig. 1.10 Effective strategy of regulation. When the external concentration is increased at point A, a switch from one transport system *1* to another *2* occurs

Fig. 1.11 Robust strategy of regulation. In the interval of concentrations between points G and F, both transport systems work simultaneously, and the internal concentration of ions remains constant and equal to $\tilde{n}^i$.

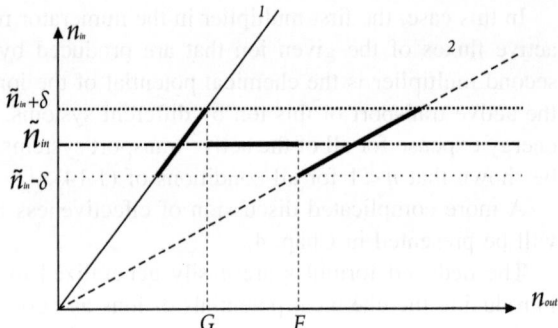

the operation frequency of one of the transport systems. Assuming that the frequency of the transport system depends on the external ion concentrations, we can derive the explicit form of this function from (1.6). For example, for C_2, we obtain the following relation:

$$C_2(n^o) = C_1 \frac{\tilde{n}^i \exp(\Delta\mu_A) - n^o}{n^o - \tilde{n}^i \exp(\Delta\mu_B)}, \qquad (1.15)$$

where $\tilde{n}^i$ is normal concentration.

For this case, the strategy has the form shown in Fig. 1.11.

Because the pump capacity can only be a positive value (in this case, the direction of the pump operation is fixed), the equality (1.15) only holds within a specific range (the range G-F in Fig. 1.11):

$$\tilde{n}^i \cdot \exp(\Delta\mu_B) < n^o < \tilde{n}^i \cdot \exp(\Delta\mu_A).$$

Above this range,

$$n^o > \tilde{n}^i \cdot \exp(\Delta\mu_A), \ C_2(n^o) = 0$$

and the internal concentration will be defined by the formula

$$n^i = n^o \cdot \exp(-\Delta\mu_A),$$

which has a linear dependence.

Below the range,

$$n^o < \tilde{n}^i \cdot \exp(\Delta\mu_B),$$

and we have correspondingly

$$n^i = n^o \cdot \exp(-\Delta\mu_B).$$

Thus, it is observed from Fig. 1.11 that in the range G-F, the system is completely robust, but the efficiency is less than 100 %. In the other ranges, the efficiency is 100 %, but the robustness is partial. It is easy to show that larger

1.3 Transport of Ions Through Cell Membranes

absolute differences between $\Delta\mu_A$ and $\Delta\mu_B$ will correspond to wider robustness ranges. However, the efficiency will still be less.

If the transport systems are assigned, it can be shown that the requirements for 100 % efficiency and complete robustness cannot be fulfilled simultaneously in either range.

1.3.6 An Optimization of the Transport System of a Cell as a Game Problem

Thus, the problem is similar to a game problem (Melkikh and Sutormina 2011). In game theory, some problems involve only one player, while nature acts as the second player. In other words, the answers are determined by the known laws of physics. In our case, the first player is a cell, which can select different strategies; that is, it can switch ion transport systems on and off. Therefore, some internal ion concentrations are established following the laws of transport phenomena. A gain (a loss) is decrements in the cell as its internal parameters deviate from the normal value. In terms of game theory, the operation of one transport system for one ion corresponds to a pure strategy, whereas the operation of two transport systems corresponds to a mixed strategy.

Considering what has been said above, it is possible to propose an algorithm for solving the problem of the regulation of ion concentrations in a cell:

1. Determine the basic transport system for a given ion (it provides the best fit to the normal value found in experiments).
2. Determine the regulatory systems from the experimental data.
3. Plot the internal versus external concentration dependence for the basic system (taking into account a possible variation of the potential).
4. Plot the internal versus external concentration dependence for the regulatory systems (taking into account a possible variation of the potential).
5. Use a graphical (or analytical) method and select the curve that most closely approaches the normal value as the external concentration changes.
6. Determine the point where it is beneficial to switch from the basic system to a regulatory system.

Write the matrix of this game:

$$A = \begin{pmatrix} a_{11} & \cdots & a_{1n} \\ \cdots & \cdots & \cdots \\ a_{m1} & \cdots & a_{mn} \end{pmatrix},$$

where m is the number of strategies of the cell, n is a number of "strategies" of the environment, and a_{ik} is the gain received by the cell. Here, under the strategies of the environment, the concentration of ions in the environment (e.g. "low" or "large") can be understood. Under the strategies of the cell, the use of a transport system for the transfer of ions can be understood. The concept of "winning" in this case is closely related to the concept of "efficiency", which is discussed above. The gain in the limiting cases, i.e. when each ion is transferred by only one system and when the concentration of ions inside the cell is maintained at a constant level, can be determined relatively easily. Generally, defining "winning" is difficult, and possible approaches for this solution will be discussed in Chap. 4.

Furthermore, when constructing models of ion transport (Chaps. 2–3), we will first use a model that is based on the considered algorithm "one ion—one transport system". Second, to simulate ion regulation in different cells, we will use the abovementioned approach. Other methods for using biological cybernetics for transport processes in cells will be discussed in Chap. 4.

Generalization of the game approach in a situation where incomplete information is present will be discussed in Chap. 4.

References

Angeli D, Sontag ED (2010) Graphs and the dynamics of biochemical networks. In: Iglesias PA, Ingalls BP (eds) Control theory in systems biology. MIT Press, Cambridge
Anokhin PK (1970) Theory of functional systems. Uspekhi Fiziol Nauk 1:19–54
Ashby WR (1954) Design for a Brain. Chapman and Hall, London
Bedau MA (2010) An Aristotelian account of minimal chemical life. Astrobiology 10(10):1011–1020
Brent R (2004) A partnership between biology and engineering. Nat Biotechnol 22(10):1211–1214. doi:10.1038/nbt1004-1211
Bustamante S, Keller D, Oster G (2001) The physics of molecular motors. Acc Chem Res 34(6):412–420
Cantone I, Marucci L, Iorio F, Ricci MA, Belcastro V, Bansal M, Santini S, di Bernardo M, di Bernardo D, Cosma MP (2009) A yeast synthetic network for In vivo assessment of reverse-engineering and modeling approaches. Cell 137(1):172–181. doi:10.1016/j.cell.2009.01.055
Caplan SR, Essig A (1983) Bioenergetics and linear nonequilibrium thermodynamics: the steady state. Harvard University Press, Cambridge
Carlson JM, Doyle J (2002) Complexity and robustness. Proc Natl Acad Sci USA 99:2538–2545
Casti JL (1979) Connectivity, complexity, and catastrophe in large-scale systems. Wiley, New York
Chiarabelli C, Stano P, Luisi PL (2009) Chemical approaches to synthetic biology. Curr Opin Biotechnol 20:492–497
Csete ME, Doyle JC (2002) Reverse engineering of biological complexity. Science 295:1664–1669
De Weer P, Gadsby DC, Rakowski RF (2001) Voltage dependence of the apparent affinity for external Na^+ of the backward-running sodium pump. J Gen Physiol 117(4):315–328
Del Vecchio D, Sontag ED (2009) Synthetic biology: a systems engineering perspective. In: Iglesias PA, Ingalls BP (eds) Control theory in systems biology. MIT Press, Cambridge
Elowitz M, Lim WA (2010) Build life to understand it. Nature 468:889–890. doi:10.1038/468889a
El-Shamad H, Goff JP, Khammash M (2002) Calcium homeostasis and parturient hypocalcemia: an integral feedback perspective. J Theor Biol 214:17–29
Faber GM, Rudy Y (2000) Action potential and contractility changes in $[Na^+]_i$ overloaded cardiac myocytes: a simulation study. Biophys J 78(5):2392–2404
Fahraeus C, Theander S, Edman A, Grampp W (2002) The K–Cl cotransporter in the lobster stretch receptor neuron—a kinetic analysis. J Theor Biol 217(3):287–309. doi:10.1006/yjtbi.3038

References

Freedman J (2001) Biophysical chemistry of physiological solutions. In: Sperelakis N (ed) Cell physiology sourcebook, 3rd edn. Academic Press, San Diego

Ganti T (2003) The principles of life. Oxford University Press, Oxford

Gibson DG, Glass JI, Lartigue C, Noskov VN, Chuang R-Y, Algire MA, Benders GA, Montague MG, Ma L, Moodie MM, Merryman C, Vashee S, Krishnakumar R, Assad-Garcia N, Andrews-Pfannkoch C, Denisova EA, Young L, Qi Z-Q, Segall-Shapiro TH, Calvey CH, Parmar PP, Hutchison CA III, Smith HO, Venter JC (2010) Creation of a bacterial cell controlled by a chemically synthesized genome. Science 329(5987):52–56. doi:10.1126/science.1190719

Goldman DE (1943) Potential, impedance, and rectification in membranes. J Gen Physiol 27(1):37–60

Haseloff J, Ajioka J (2009) Synthetic biology: history, challenges and prospects. J R Soc Interface 6(4):S389–S391. doi:10.1098/rsif.2009.0176.focus

Hodgkin AL, Horowicz P (1959) The influence of potassium and chloride ions on the membrane potential of single muscle fibers. J. Physiol 148:127–160

Hodgkin AL, Katz B (1949) The effect on sodium ions in electrical activity of the giant axon of the squid. J Physiol 108(1):37–77

Holling CS (1973) Resilience and stability of ecological systems. Annu Rev Ecol Syst 4:1–23

Hopfer U (2002) A Maxwell's Demon type of membrane transport: possibility for active transport by ABC-transporters? J Theor Biol 214(4):539–547

Iglesias PA, Ingalls BP (2009) Control theory and systems biology. MIT Press, Cambridge

Jacquez J (1972) A compartmental analysis in biology and medicine. Elsevier Pub Co., New York

Jamshidi N, Palsson BO (2006) Systems biology of the human red blood cell. Blood Cells Mol Dis 36(2):239–247

Kabakov AY (1994) The resting potential equations incorporating ionic pumps and osmotic concentrations. J Theor Biol 169(1):51–64

Kabakov AY (1998) Activation of K_{ATP} channels by Na/K pump in isolated cardiac myocytes and giant membrane patches. Biophys J 75(6):2858–2867

Kedem O, Katchalsky A (1958) Thermodynamic analysis of the permeability of biological membranes to non-electrolytes. Biochim Biophys Acta 27(2):229–246

Kitano H (2004) Biological robustness. Nat Rev Genet 5:826–837

Kitano H (2007) Towards a theory of biological robustness. Mol Syst Biol 3:137. doi:10.1038/msb4100179

Kjelstrup S, Rubi JM, Bedeaux D (2005) Active transport: a kinetic description based on thermodynamic grounds. J Theor Biol 234(1):7–12

Melkikh AV, Seleznev VD (2005) Models of active transport of ions in biomembranes of various types of cells. J Theor Biol 324(3):403–412

Melkikh AV, Seleznev VD (2006a) Requirements on models and models of active transport of ions in biomembranes. Bull Math Biol 68(2):385–399

Melkikh AV, Seleznev VD (2006b) Model of active transport of ions in biomembranes based on ATP-dependent change of height of diffusion barriers to ions. J Theor Biol 242(3):617–626

Melkikh AV, Seleznev VD (2008) Early stages of the evolution of life: a cybernetic approach. Orig Life Evol Biosph 38(4):343–353

Melkikh AV, Seleznev VD (2012) Mechanisms and models of the active transport of ions and the transformation of energy in intracellular compartments. Prog Biophys Mol Biol 109(1–2):33–57

Melkikh AV, Sutormina MI (2008) Model of active transport of ions in cardiac cell. J Theor Biol 252(2):247–254

Melkikh AV, Sutormina MI (2011) Algorithms for optimization of the transport system in living and artificial cells. Syst Synth Biol 5:87–96. doi:10.1007/s11693-011-9083-6

Munteanu A, Sole RV (2006) Phenotypic diversity and chaos in a minimal cell model. J Theor Biol 240:434–442

Murdoch WM (1973) Population regulation and population inertia. Ecology 51:497–502

Murtas G (2007) Question 7: construction of a semi-synthetic minimal cell: a model for early living cells. Orig Life Evol Biosp 37:419–422

Murtas G (2009) Artificial assembly of a minimal cell. Mol Biosyst 5(11):1292–1297

Novoseltsev VN (1978) Teoriya upravleniya i biosistemy (control theory and biosystems) control theory and biosystems. Nauka, Moscow

Rashevsky N (1960) Mathematical biophysics : physico-mathematical foundations of biology, 3d edn. Dover Publications, New York

Rasmussen S, Chen L, Stadler BMR, Stadler PF (2004) Proto-organism kinetics: evolutionary dynamics of lipid aggregates with genes and metabolism. Orig Life Evol Biosp 34:171–180

Sagar A, Rakowski RF (1994) Access channel model for the voltage dependence of the forward-running Na^+/K^+ pump. J Gen Physiol 103(5):869–893

Senachak J, Vestergaard M, Vestergaard R (2007) Cascaded games. Lect Notes Comput Sc 4545:185–201

Shoemaker JE, Chang PS, Kwei EC, Taylor SR, Doyle FJ III (2010) Robustness and sensitivity analyses in cellular networks. In: Iglesias PA, Ingalls BP (eds) Control theory in systems biology. MIT Press, Cambridge

Slepchenko BM, Schaff JC, Carson JH, Loew LM (2002) Computational cell biology: spatiotemporal simulation of cellular events. Annu Rev Biophys Biomol Struct 31:423–441

Smolke CD, Silver PA (2011) Informing biological design by integration of systems and synthetic biology. Cell 144(6):855–859

Solé RV, Munteanu A, Rodriguez-Caso C, Macía J (2007) Synthetic protocell biology: from reproduction to computation. Philos Trans R Soc Lond B Biol Sci 362:1727–1739

Sperelakis N (2001) Origin of resting membrane potential. In: Sperelakis N (ed) Cell physiology sourcebook, 3rd edn. Academic Press, San Diego

Stelling J, Sauer U, Szallasi Z, Doyle FJ, Doyle J (2004) Robustness of cellular functions. Cell 118:675–685

Tsong TY, Chang CH (2003) Ion pump as brownian motor: theory of electroconformational coupling and proof of ratchet mechanism for Na–K–ATPase action. Physica A 321(1–2):124–138

von Bertalanffy L (1973) General system theory: foundations, development, applications. Penguin, London

Waterman TH (1968) Systems theory and biology—view of a biologist. In: Mesarovic MD (ed) Systems theory and biology. Springer, New York

Wiener N (1961) Cybernetics: or control and communication in the animal and the machine, 2nd edn. Wiley, New York

Zerges W (2002) Does complexity constrain organelle evolution? Trends Plant Sci 7(4):175–182

Chapter 2
Models of Ion Transport in Mammalian Cells

Cardiomyocytes, neurons, hepatocytes and erythrocytes are considered based on the algorithm "one ion—one transport system" models of some mammalian cells. Models of the compartments of mammalian cells, e.g., synaptic vesicles, sarco- and endoplasmic reticulum and mitochondria, are built; and models for the regulation of ion transport in mammalian cells and their compartments are presented. We find conditions under which a robust and effective strategy for the switching of transport systems of cells takes place.

2.1 Introduction

The basis of cellular forms of life lies in a process that involves continuous maintenance of the differential concentrations of ions inside the cell relative to that of the extracellular medium. Thus, most mammalian cells, when compared to the extracellular medium, have low concentrations of sodium and calcium and large concentrations of potassium ions. The constancy of these differences is critical for homeostasis and plays an important role in the regulation of metabolism, and this distribution is due to membrane permeability and ion pumps that are built into those membranes. Because of passive permeability, the cell has a tendency to equalize concentrations; however, pumps work to maintain a concentration gradient. Pump operation requires energy, in the form of ATP, which is generated by the combustion of the "fuel" of the cell, especially fats and carbohydrates, in the mitochondria. ATP has a large free energy ($\sim$ 150 kJ/mol) and is transported to the membrane pumps, which function constantly.

Molecules of transferred substances or ions can be transported by various manners: they can travel through the membrane, irrespective of the transfer of other compounds (uniport), their transfer can be carried out simultaneously and unidirectionally to other compounds (symport), or transport connections can be simultaneously and oppositely directed (antiport). Symport and antiport are systems of co-transport for which the rate of the overall process is controlled by the availability and accessibility of transport systems for both ions. In turn, co-transport is a type active transport. Passive and active transport are provided with special structures such as channels, transporters, and enzymes that ensure the movement of specific ions against their concentration gradients, and this action is dependent upon the energy of ATP. This transfer is carried out by transport ATPases, which are also referred to as ion pumps.

In this chapter, we will build transport system models for some cells based on the proposed algorithm in Chap. 1. To validate the algorithm, the model must be applied to cells—the representatives of the various kingdoms of nature, and we use the classification of the animal world, which was offered by Cavalier-Smith. In this work (Cavalier-Smith 2002; Cavalier-Smith 1998), all life was divided into two domains: eukaryotes and prokaryotes. Eukaryotes are divided into five kingdoms: animals, plants, fungi, protists and chromists. Prokaryotes were classified into archaea and eubacteria.

The considered mammalian cells possess a number of features that differentiate them from cells of other kingdoms. First, the ionic composition of their external environment is nearly constant, and second, the lack of flexibility of their cell walls and membranes require a minimum differential pressure across the membrane (i.e., the difference of internal and external pressures).

In addition to the four types of mammalian cells in this chapter, models of transport systems will be built for the mitochondria, synaptic vesicles, and the sarco- and endoplasmic reticula. By ignoring features of these compartments that are irrelevant for our algorithm, we consider these compartments as isolated systems that possess the ability to maintain intracellular concentrations at a given level.

2.2 Cardiac Cells

Functioning of a cardiac muscle cell considerably depends on the ion concentrations in the cell and its environment. These ions primarily include Na^+, K^+, Ca^{2+}, Mg^{2+}, HCO_3^- and Cl^-. For example, calcium ions participate in the contraction of muscles and signal processes, while magnesium ions specifically control the work of ATP-dependent exchangers. Intracellular magnesium is extremely important for intactness and the functioning of ribosomes, and it takes part in regulating the concentrations and transport of calcium, potassium, sodium and phosphate ions inside and outside the cell. As a cofactor, magnesium activates over 300 enzymatic reactions that are involved in metabolic processes in organisms. Magnesium interacts with cellular lipids, ensures the intactness of the cell membrane, and enters a competitive relationship with calcium for contraction elements. Furthermore, the concentrations of calcium, potassium and sodium ions largely determine the properties of the action potential in a cardiac muscle cell.

2.2 Cardiac Cells

Table 2.1 Experimental data for ion concentrations and the resting potential for a cardiac cell

Ions	Internal concentrations, mM	External concentrations, mM
Na^+	12	145
K^+	155	4
Mg^{2+}	$0.5 \div 1$	$1 \div 2$
Ca^{2+}	$10^{-3} \div 10^{-4}$	1
Bicarbonate	8	27
Cl^-	4	120
Others	155	7
Potential, mV	$-(83 \div 100)$	

All of the aforementioned processes are strongly sensitive to intracellular concentrations of Ca^{+2} and Mg^{+2} ions. These concentrations can change depending on the concentrations of the corresponding ions in the cellular environment. However, models for the independent prediction of ion concentrations in a cell and the membrane resting potential are unavailable in the current literature.

Details of the construction of the transport system for cardiomyocytes in the steady state have been established (Melkikh and Sutormina 2008). Here, we briefly present the main results, and based on the model of the transport system, we consider algorithms for the regulation of ion transport.

First, information on the values of the concentrations of internal and external ions are needed, and Table 2.1 gives the concentrations of the basic types of ions that are found inside and outside of a cell (Sperelakis 2000; Murphy 2000; Sperelakis and Gonzales-Serratos 2001).

The main and most well-studied pump in cardiac muscle cells is the Na^+–K^+-ATPase. This pump is ATP-dependent, and it transfers three sodium ions from the cell and exchanges them for two potassium ions. In addition to this pump, we will consider other sodium ion exchangers. For example, we will discuss the Ca^{2+}–Na^+ exchanger, which transports one calcium ion in exchange for three sodium ions, and the Mg^{2+}–Na^+ exchanger, which transports one magnesium ion in exchange for one sodium ion. Additionally, electrically neutral co-transport transports a sodium ion, potassium ion and two chlorine ions in one direction—K^+–Na^+–$2Cl^-$. In turn, calcium ions are pumped out of the cell by the Ca^{2+}-ATPase. Additionally, all ions can passively penetrate the through membrane, the magnitude of passive flow is determined by the membrane permeability coefficient P for each type of ion. Other mechanisms exist to transport chloride ions in addition to passive flow and the K^+–Na^+–$2Cl^-$ co-transporter such as the K^+–Cl^- and Na^+–Cl^- symporters, which transfer potassium and chlorine or sodium and chlorine out of the cell, respectively, and the OH^-Cl^- antiporter, which transports chlorine through the membrane into the cell and the HCO—Cl—H^+–Na^+ co-transporter, which exchanges bicarbonate and sodium ions for a chloride ion and proton (Luo and Rudy 1991, 1994; Faber and Rudy 2000; Hume et al. 2000).

All of the described mechanisms are shown in Fig. 2.1.

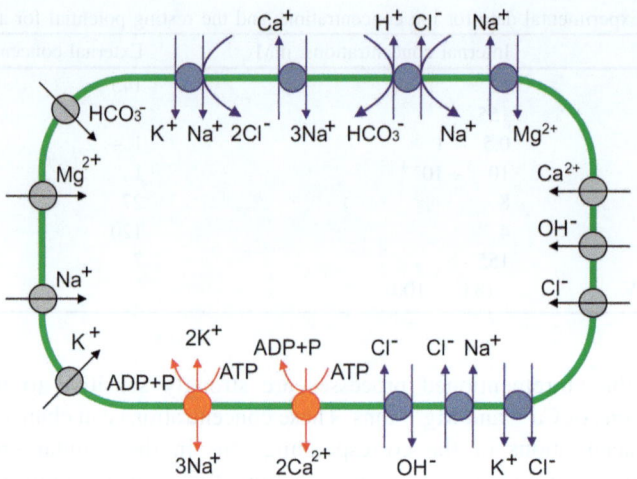

Fig. 2.1 Transport systems in a cardiac cell

2.2.1 Model of Transport Systems

Sodium ions are carried by several transport systems (Sperelakis 2000), but the main pump that transfers this ion to the external environment of the majority of animal cells is thought to be the Na$^+$–K$^+$-ATPase. Despite the fact that other transport mechanisms can provide a smaller deviation between the required and internal concentrations of sodium ions, the effect of ATPase activity is necessary to create a non-equilibrium state in the system. Thus, we chose this system as the basis for sodium ions. The Na$^+$–K$^+$-ATPase carries three sodium ions out of the cell and two potassium ions into the cell. In accordance with the earlier proposed model, the equation for the sodium flow that is produced by this system has the following form:

$$J_{Na-K-ATP} = 3C_{Na-K-ATP}\left[\exp(\Delta\mu_A + \varphi)(n_{Na}^i)^3(n_K^o)^2 - (n_{Na}^o)^3(n_K^i)^2\right], \quad (2.1)$$

where $C_{Na-K-ATP}$ is the constant of the active transport of sodium and potassium ions, and the remaining symbols have been given previously.

The analytical expression for the dependence of the internal concentrations of sodium ions on the values of the membrane potential and extracellular concentrations can be obtained after we determine the main transport system for potassium ions. The following systems transport potassium ions:

- Na$^+$–K$^+$-ATPase;
- Na$^+$–K$^+$–2Cl$^-$—co-transport;
- K$^+$–Cl$^-$—co-transport;
- Potassium channels.

2.2 Cardiac Cells

Because of the result from the calculation of the internal concentration of potassium ions that are produced by each of these systems and its comparison with the experimental data, it was concluded that the distribution of potassium ions is near the Boltzmann distribution because they can easily penetrate through the biomembrane.

We shall neglect the active flows of potassium ions relative to their passive flow; therefore, their concentration follows the Boltzmann distribution:

$$n_K^i = n_K^o \exp(-\varphi), \tag{2.2}$$

From (2.1) and (2.2), the expression that models the dependence of the internal parameters on the external parameters for sodium ions is as follows:

$$n_{Na}^i = n_{Na}^o \cdot \exp\left(-\varphi - \frac{\Delta\mu_A}{3}\right) \tag{2.3}$$

When considering chlorine transport as the third potential-forming type of ion transport, we found the minimal deviation from the actual values that were provided by the K–Cl, Na–Cl–HCO$_3$–H co-transporters and the passive penetration of chloride ions through their electrochemical gradient. Notably, the passive distribution of potassium ions, which is the dependence of the internal concentration of chlorine on the external concentration of K–Cl, is also described by the Boltzmann equation. The calculated value that is created by a complex mechanism of Na–Cl–HCO$_3$–H is near that of the calculated value of the K–Cl co-transporter and passive penetration of chloride ions, and it can be assumed that the analytical expression that depends on the internal concentration of the external parameters for this type of transport system will be similar. Therefore, the concentration of these anions in a cell approaches the Boltzmann distribution (although chlorine is also involved in several transport systems (Sperelakis 2000; Hume et al. 2000)):

$$n_{Cl}^i = n_{Cl}^o \exp(\varphi). \tag{2.4}$$

The focus of a work by Melkikh and Sutormina (2008) was given to the transport of divalent cations, which play a special role in carrying out vital functions in heart muscle cells.

In addition to passive transport across the cell membrane through the electrochemical potential, calcium in a cardiac muscle cell is transported by two systems: the Ca-ATPase and a Na$^+$–Ca^{2+}-exchanger.

It is known that a Na$^+$–Ca^{2+}-exchanger exchanger substitutes one calcium ion for three sodium ions in the absence of the consumption of ATP energy. Thus, in accordance with the model (Melkikh and Sutormina 2008), the equation has the following form:

$$J_{Ca-Na} = C_{Ca-Na} \cdot \left[n_{Ca}^i \cdot (n_{Na}^o)^3 - \exp(\varphi) n_{Ca}^o \cdot (n_{Na}^i)^3\right], \tag{2.5}$$

The structure of the equation reflects the exchanger stoichiometry (i.e., one calcium ion is exchanged for three sodium ions) and the fact that sodium and calcium ions are transported in opposite directions.

Because one calcium ion is actively transported out of a cell during the hydrolysis of one ATP molecule, we obtain an equation for the active transport of calcium, which is in line with that of (Melkikh and Sutormina 2008):

$$J_{Ca-ATP} = C_{Ca-ATP}\left[\exp(\Delta\mu_A + 2\varphi)n_{Ca}^i - n_{Ca}^o\right], \qquad (2.6)$$

The calculated values of the intracellular concentration of Ca ions for each system are the following:

- Ca–ATФ: $n_{Ca}^i = 2.7 \cdot 10^{-6}$ mM;
- Na–Ca: $n_{Ca}^i = 1.5 \cdot 10^{-5}$ mM;
- Passive flux: $n_{Ca}^i = 1.3 \cdot 10^3$ mM.

When comparing these values with those from the experimental data ($10^{-3} \div 10^{-4}$), we see that the sodium exchanger produces the smallest deviation. However, the order of the values of the intracellular concentration of calcium ions is small when compared to the concentrations of other ions. In other words, in the first approximation in our algorithm, we can use an ATP-dependent calcium pump and exchanger. This situation is similar to the choice of the main system for sodium and we assume for the main ATP-dependent pump. Thus, the dependence of the internal concentration of calcium ions on the external concentrations (2.6) takes the following form:

$$n_{Ca}^i = n_{Ca}^o \cdot \exp(-2\varphi - \Delta\mu_A) \qquad (2.7)$$

Transport of another divalent cation, magnesium, in cardiomyocytes is provided by the following mechanisms (Freudenrich et al. 1992; Murphy 2000):

- Na^+–Mg^{2+}-exchanger;
- Magnesium channels.

Expressions of fluxes that are generated by these transport systems can be written in the following form:

$$J_{Na-Mg} = C_{Na-Mg} \cdot \left[\exp(\varphi)n_{Mg}^{in} \cdot n_{Na}^{out} - n_{Mg}^{out} \cdot n_{Na}^{in}\right] \qquad (2.8)$$

$$J_{Mgpas} = P_{Mg} \cdot \left[n_{Mg}^{in} \cdot \exp(2\varphi) - n_{Mg}^{out}\right] \qquad (2.9)$$

Numerical calculations showed that passive flow creates an internal concentration of 2.7×10^3 mM, while the counter-transport of sodium and magnesium provides a value of 6 mM, which is in better agreement with the experimental data ($0.5 \div 1$ mM). Therefore, the expression for the intracellular concentration of magnesium ions, in view of (2.3), can be written as the following:

2.2 Cardiac Cells

$$n^i_{Mg} = n^o_{Mg} \cdot \exp\left(-\frac{\Delta\mu_A}{3} - 2\varphi\right) \quad (2.10)$$

The next anion transport that will be considered to simulate ion transport in cardiomyocytes is the bicarbonate ion, HCO_3. The transfer of this ion is provided by work of the electroneutral complex Na–Cl–HCO_3–H, chlorine-bicarbonate exchanger and through special channels. The $Na^+ - HCO_3^- - Cl^- - H^+$ co-transporter operates as follows: a sodium and bicarbonate ion is exchanged for a chlorine anion and a proton, and this mechanism is electroneutral. Because the cell potential does not produce a barrier, the work of the exchanger depends on the concentration of ions and the probability that they will interact with the carrier. Considering the earlier proposed model, we can write the following equation:

$$J_{Na-HCO_3^--Cl-H} = C_{Na-HCO_3^--Cl-H}(n^i_{HCO_3^-} \cdot n^i_{Na} \cdot n^o_{Cl} \cdot n^o_H - n^o_{HCO_3^-} \cdot n^o_{Na} \cdot n^i_{Cl} \cdot n^i_H), \quad (2.11)$$

where $C_{Na-HCO_3^--Cl-H}$ is the exchange constant for the HCO_3^-, Cl^-, Na^+ and H^+ anions.

Considering Eq. (2.11), the dependence of the intracellular concentration of the anions on their extracellular concentration and the membrane potential can be written as follows:

$$n^i_{HCO_3^-} = n^o_{HCO_3^-} \exp\left(\frac{\Delta\mu_A}{3} + 2\varphi\right). \quad (2.12)$$

Substituting the known values results in $n^i_{HCO_3^-} = 13.9$ mM,

The expression for the flow of ions that are produced by the $HCO_3^- - Cl^-$ exchanger can be written as follows:

$$J_{HCO_3^--Cl^-} = C_{HCO_3^--Cl^-}(n^i_{HCO_3^-} \cdot n^o_{Cl} - n^o_{HCO_3^-} \cdot n^i_{Cl}). \quad (2.13)$$

It is observed from this expression that the operation of this exchanger is equivalent to the passive transport of HCO_3^- ions. When the known values are substituted, $n^i_{HCO_3^-} = 0.7$ mM, and by equating the flow to zero, we obtain the following expression for the intracellular concentration of HCO_3^- ions when only this exchanger operates:

$$n^i_{HCO_3^-} = n^o_{HCO_3^-} \exp(\varphi).$$

Thus, we may conclude that the main transport system for HCO_3^- ions is the sodium-dependent exchanger because the intracellular concentration of HCO_3^- ions that are produced by this exchanger most closely approaches the experimental value.

However, this conclusion does not suggest that the role of the $HCO_3^- - Cl^-$ exchanger is insignificant in a cell. This exchanger can be significant in regulating the concentration of HCO_3^- ions and pH inside of the cell.

We decided on the main transport system for each type of ion. According to our algorithm and by using the equation of electrical neutrality, we derive the dependence of the membrane resting potential on the extracellular concentrations of ions that are under consideration:

$$2 \cdot n_{Ca}^o \cdot \exp(-\Delta\mu_A - 2\varphi) + n_{Na}^o \cdot \exp\left(-\frac{\Delta\mu_A}{3} - \varphi\right) + n_K^o \cdot \exp(-\varphi) +$$
$$+ 2 \cdot n_{Mg}^o \cdot \exp\left(-\frac{\Delta\mu_A}{3} - 2\varphi\right) = n_{Cl}^o \cdot \exp(\varphi) + 0.89 \cdot n_{HCO}^o \exp\left(2\varphi + \frac{\Delta\mu_A}{3}\right) + Z_A \cdot n_A \quad (2.14)$$

Concentration $Z_A \cdot n_A$ of non-penetrating ions is known experimentally. By solving the equation numerically, it can be shown that the potential is equal to -3.7 (approximately -92.5 mV), which closely corresponds to the experimental data.

From Eq. (2.14), we can explicitly obtain an expression for the potential. Because the contributions of calcium, magnesium, and bicarbonate in the potential are small in comparison with other ions, they will be neglected; therefore, the dependence of the membrane potential on the extracellular concentrations has the following form:

$$\varphi = -\ln\left(\frac{\sqrt{\left(\frac{Z_A n_A}{2n_{Cl}^{out}}\right)^2 + \left(\frac{n_{Na}^{out}}{n_{Cl}^o}\exp\left(\frac{-\Delta\mu_A}{3}\right) + \frac{n_K^{out}}{n_{Cl}^{out}}\right)} + \frac{Z_A n_A}{2n_{Cl}^{out}}}{\left(\frac{n_{Na}^{out}}{n_{Cl}^{out}}\exp\left(\frac{-\Delta\mu_A}{3}\right) + \frac{n_{Na}^{out}}{n_{Cl}^{out}}\right)}\right) \quad (2.15)$$

Graphically, the dependence of the membrane potential on changes in the extracellular medium is shown on Fig. 2.2.

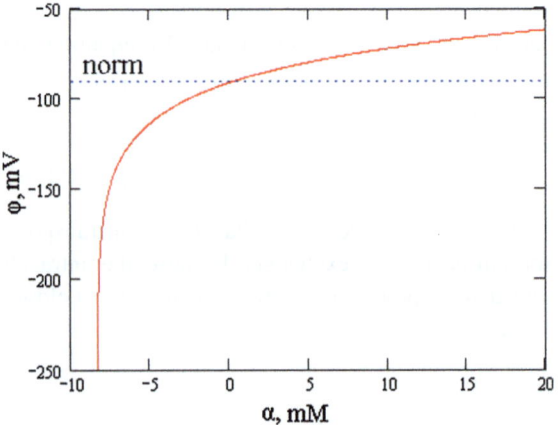

Fig. 2.2 Dependence of the potential on the addition of potassium and chlorine ions to the environment

2.2 Cardiac Cells

Table 2.2 Comparisons of the calculated values of the internal ion concentrations and membrane potential with the experimental data from cardiomyocytes

Ions	Experimental data, mM	Calculated data, mM
Sodium	12	6.9
Potassium	155	151.2
Magnesium	$0.5 \div 1$	3.6
Calcium	$10^{-3} \div 10^{-4}$	2.9×10^{-6}
Bicarbonate	8	13.2
Chlorine	4	3.1
Potential, mV	$-(83 \div 100)$	-90.8

In the Fig. 2.2, the solid line indicates the change in the resting membrane potential after adding a certain concentration of α (in mM) into the extracellular medium in an equal amount to that of potassium and sodium ions with chlorine. The points show the normal value of the membrane potential.

The value of the membrane potential without divalent cations is slightly smaller than the absolute value of approximately -90.8 mV, which is also consistent with the experimental data. By including (2.15), i.e., dependencies of the intracellular on the external concentration, we can calculate the values of the internal concentrations of ions, and detailed results of the calculation are given in (Melkikh and Sutormina 2008). The obtained results are tabulated in Table 2.2.

2.2.2 Regulation of Ion Transport

We have constructed a system of equations that describe the transport processes in the cell during the stationary state, and these equations can be used as a starting point for modeling the regulation of ion transport through the membrane under varying external conditions.

As mentioned earlier, potassium ion transport is carried out by four methods in cardiomyocytes: an ATP-dependent pump, two co-transporters and passive transport. According to our algorithm, we can construct the dependence of the intracellular on the extracellular concentration that is produced independently by each mechanism. We can then graphically determine an effective and robust strategy for regulation. The potential changes are taken into account when constructing strategies for the regulation of ions, which strongly affect the potential.

The work of the four transport systems are represented in the following graph (Fig. 2.3).

The figure displays the dependence of the intracellular concentration of potassium ions from the addition of α on the external environment of potassium ions with chlorine, in mM. The solid line indicates passive flow, points represent Na–K–Cl-co-transport, the dashed line denotes the actual concentration of potassium in the cell at steady state, and the dash-dot line represents the work of Na–K-ATP.

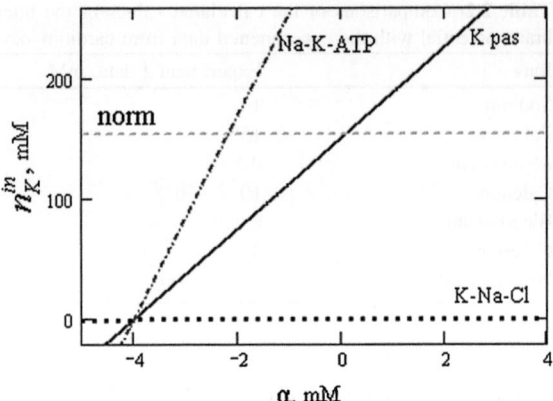

Fig. 2.3 Mechanisms for the transfer of potassium ions in cardiac cells

The figure shows that the dependence of the internal on external concentrations of ions is linear; however, this finding does not indicate that the potential depends weakly on the concentration of extracellular potassium. As shown in the figure, in the selected range, only a partial dependency is reflected, and the interval changes in the external concentration is limited to values that are approximately twice the normal value because strong deviations lead to cell death.

By using the algorithm for finding the optimal strategy for transport, the optimal strategy for the transport system of potassium ions can be determined graphically.

We show the selected, best strategy for transport systems of cardiac muscle cells in Fig. 2.4.

In the figure, the solid line indicates passive flow, the points refer to Na–K–Cl-co-transport, the dashed line denotes the actual concentration of potassium in the cell at steady state, the dash-dot line refers to the work of the Na–K-ATPase, and the bold solid line indicates the chosen strategy.

We utilize the optimal strategy to ensure the maximum efficiency of pumps and exchangers at any given time, i.e., only one transport mechanism is functional.

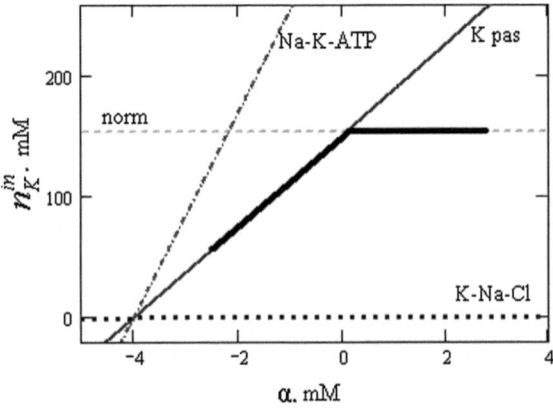

Fig. 2.4 Strategy for regulating potassium ion transport in cardiomyocytes

2.2 Cardiac Cells

However, as observed in Fig. 2.4, a significant increase in the external concentration ensures that the robustness of cells that is required for simultaneous operation of at least two mechanisms occurs. From the literature, it is known that most potassium ions are transferred passively, but when the extracellular concentration of this ion is doubled, the electrically neutral Na–K–Cl co-transporter becomes functional.

To account for changes in two parameters, one can construct three-dimensional plots of the intracellular concentration of one ion in relation to changes in the external concentrations of the two types of ions. A change in potential will also affect the internal concentration of the desired value because it appears in the expression for determining the intrinsic value of the concentration, and it is an explicit function of external concentrations. When required to consider an effective strategy (i.e., at each moment, only one transport mechanism for each type of ion is present), the strategy will be a set of points on surfaces, which indicate the mechanisms of ion transport, and in this case, we introduce an objective function. Because changing variables (e.g., intracellular concentration and resting potential) differ substantially in value (sometimes by orders of magnitude) and have different dimensions (e.g., concentrations and potential), it is convenient to choose the following as the objective function:

$$\Delta = \sqrt{\sum_i \left(\frac{\Delta n_i}{n_i}\right)^2 + \left(\frac{\Delta \varphi}{\varphi}\right)^2} \to \min. \quad (2.16)$$

We will use such a criterion when changing the concentrations of two or more ions.

The following is a plot of the work of the transport systems in cardiomyocytes as a function of changes in the extracellular concentrations of potassium and sodium ions and taking into account the change in potential (Fig. 2.5).

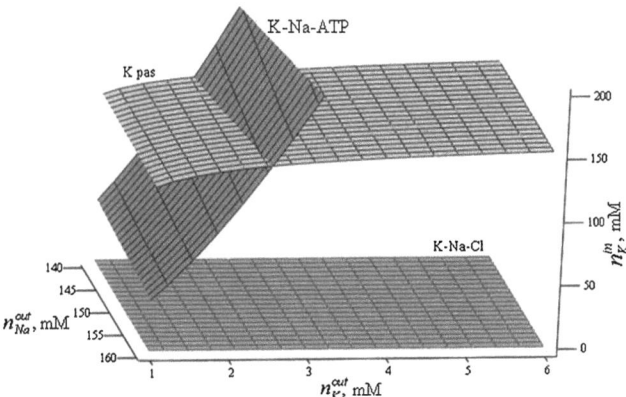

Fig. 2.5 Mechanisms of transport of potassium ions as functions of the external concentrations of two ions

It is evident from the figure that although the values deviate from those that are required for potassium ions, these ions remain passively transported. It may be noted that the ATP-dependent pump allows the required level of internal concentration values to be achieved for different values of the external concentration of sodium ions when the extracellular medium has a low content of potassium ions. However, it is only possible to maintain the normal value by passive transfer of ions through their electrochemical gradient. If a significant increase in the concentration of potassium ions outside the cell occurs, the internal concentration must exceed the normal value to maintain the desired value, and this process should include the K–Na–Cl-co-transporter. It should be noted that when simultaneously including several types of ions in the description and at a changing of two parameters, this picture could change.

2.3 Neurons

The electrical properties of neurons of different organisms that are at rest (i.e., in the absence of nerve impulses) are very similar. The main difference lies in that the external environment of neurons may differ in composition. We distinguish two different examples: mammalian neurons and neurons of marine organisms (e.g., a well-studied neuron of the squid). In the second case, the composition of the fluid surrounding the neuron is near the composition of sea water, which contains significantly more sodium, chloride and magnesium than in the blood of mammals.

Tables 2.3 and 2.4 list the ion concentrations in the neurons of squid and mammals (Raupach and Ballanyi 2004; El-Mallakh 2004; Barish 1991), correspondingly.

In addition to passive transport for all type of ions, neurons contain each of the following types of systems for ion transport (Craciun et al. 2005; Nicholls et al. 2001; Giffard et al. 2000; Chesler 2003):

- Na^+–K^+-ATPase, Ca^{2+}-ATP-ase;
- Co-transporters: Na^+–K^+–Cl^-, Na^+–Cl^-, Na^+–HCO_3^-, K^+–Cl^-;
- Exchangers: Na^+–Mg^{2+}, Na^+–H^+, Na^+–Ca^{2+}, Cl^-–HCO_3^-.

Table 2.3 Experimental data on the concentrations of ions for the neurons of squid

Ions	Internal concentrations, mM	External concentrations, mM
Potassium	360	10
Sodium	69	425
Chlorine	157	496
Calcium	0.0001	10
Magnesium	10	50
Bicarbonate	8	25

2.3 Neurons

Table 2.4 Experimental data on the concentrations of ions for the neurons of mammals

Ions	Internal concentrations, mM	External concentrations, mM
Potassium	150	5.5
Sodium	15	150
Chlorine	9	125
Calcium	0.0001	2.5
Magnesium	0.7	1
Bicarbonate	8	25

Value of the membrane potential: $(65 \div 70)$ mV

Figure 2.6 shows the major ion transport systems in neurons:

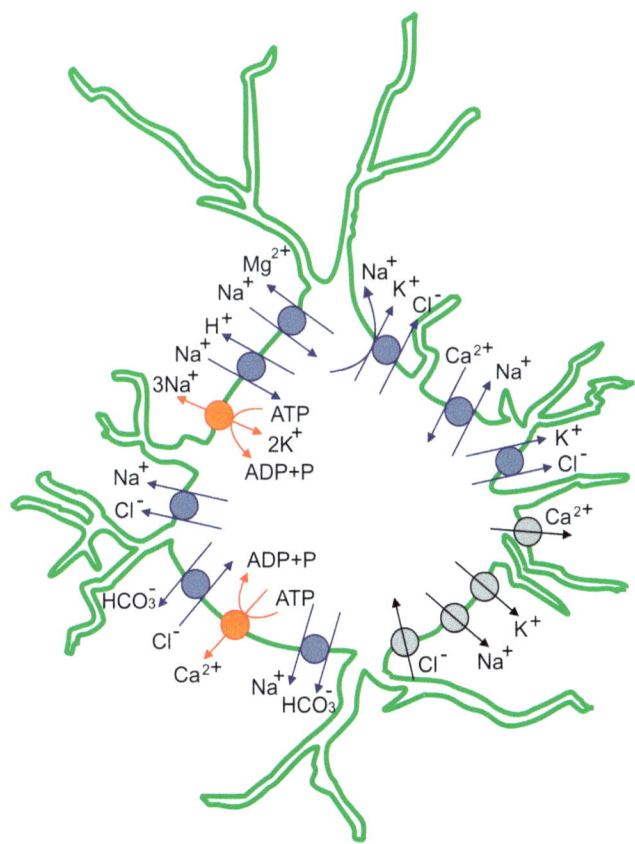

Fig. 2.6 Transport systems in neurons

2.3.1 Model of Transport Systems

Based on the model of active transport of ions that was proposed in the first chapter, the different transport systems for all types of ions are analyzed, the basic system is determined, and the flow of ions is recorded.

The transport of potential-forming ions is first modeled. According to the algorithm, we assume that the sodium ions are transported mainly by the Na–K-ATPase. Potassium is transported by a number of systems: passive transport, the sodium–potassium-chlorine co-transporter, the sodium–potassium pump, or the potassium-chlorine co-transporter. The expressions for the flows that are generated by these systems are similar to what was recorded for cardiomyocytes. Using the known value for the membrane potential and extracellular ion concentrations, the intracellular concentrations of potassium ions are calculated to be the following:

- for the neurons of squid:
    - Na–K-ATP: $n_K^i = 3.9 \cdot 10^3$ mM;
    - K–Na–Cl: $n_K^i = 615$ mM;
    - K–Cl: $n_K^i = 31.6$ mM;
    - Passive flux: $n_K^i = 134.6$ mM.

- for the neurons of mammals:
    - Na–K-ATP: $n_K^i = 1.01 \cdot 10^3$ mM;
    - K–Na–Cl: $n_K^i = 1.1 \cdot 10^4$ mM;
    - K–Cl: $n_K^i = 76.4$ mM;
    - Passive flux: $n_K^i = 74.1$ mM.

By comparing the calculated values with the known experimental data for the squid neuron, which was 360 mM, and mammalian neurons, which was 150 mM, we note that the minimum deviation results in a passive flow of ions through the membrane. We find that the dependence of the internal concentration of potassium ions and potential on external concentrations will be similar to that which was described for cardiac muscle cells. Therefore, the expression that models the dependence of the internal concentration of sodium ions of neurons will also be analogous to that for the cardiac cell.

Chlorine ions are transferred by the following systems: sodium-chlorine co-transporter, sodium–potassium-chlorine co-transporter, chlorine-bicarbonate exchanger, potassium-chlorine co-transporter, and passive flow of ions through the membrane. The expressions for the flow of chlorine ions that is produced by these systems are similar to those that are recorded in item 2.1. By using the known value of the membrane potential and ion concentrations, the intracellular concentration of chlorine is calculated as the following:

- for neurons of squid (the experimental value for the concentration of chlorine ions is 157 мM):

2.3 Neurons

- K–Na–Cl: $n^i_{Cl} = 205.1\,\text{mM}$;
- Na–Cl: $n^i_{Cl} = 3.1 \cdot 10^3\,\text{mM}$;
- K–Cl: $n^i_{Cl} = 13.8\,\text{mM}$;
- Cl–HCO$_3$: $n^i_{Cl} = 158.7\,\text{mM}$;
- Passive flux: $n^i_{Cl} = 36.8\,\text{mM}$.

- for neurons of mammals (the experimental value for the concentration of chlorine ions is 9 mM):

  - K–Na–Cl: $n^i_{Cl} = 75.7\,\text{mM}$;
  - Na–Cl: $n^i_{Cl} = 1.2 \cdot 10^3\,\text{mM}$;
  - K–Cl: $n^i_{Cl} = 4.6\,\text{mM}$;
  - Cl–HCO$_3$: $n^i_{Cl} = 40\,mM$;
  - Passive flux: $n^i_{Cl} = 9.3\,\text{mM}$.

For the squid neuron, the minimum deviation from the experimental values provides the electro-neutral exchanger with bicarbonate ions. For mammalian neurons, passive penetration of chloride ions through their electrochemical gradient occurs, thus, the dependence of the internal values of this type of ion on the external values will have the form (2.14). The expression for the squid neuron can be written after the main transport system for HCO$_3$ is found.

In the literature, the transport of bicarbonate by a chlorine-bicarbonate exchanger, passive transport and sodium-bicarbonate co-transporter (Craciun et al. 2005) has been verified in neurons. Expressions of the fluxes that are created by the first two transport systems are shown above, and the formula for the flow that is generated by the third co-transporter is the following:

$$J_{Na-HCO_3} = C_{Na-HCO_3} \cdot [(n^i_{HCO_3})^3 n^i_{Na} - \exp(2\varphi)(n^o_{HCO_3})^3 n^o_{Na}] \qquad (2.17)$$

The values for the internal concentrations of HCO$_3$, which are provided by each mechanism, can be calculated as follows:

- for the neurons of squid:

  - Cl–HCO$_3$: $n^i_{HCO_3} = 7.9\,\text{mM}$;
  - Na–HCO$_3$: $n^i_{HCO_3} = 8.1\,\text{mM}$;
  - Passive flux: $n^i_{HCO_3} = 1.9\,\text{mM}$.

- for mammalian neurons:

  - Cl-HCO$_3$: $n^i_{HCO_3} = 1.8\,\text{mM}$;
  - Na–HCO$_3$: $n^i_{HCO_3} = 9.5\,\text{mM}$;
  - Passive flux: $n^i_{HCO_3} = 1.9\,\text{mM}$.

The experimental data (8 mM for both neurons) best fits the value that is created by the sodium-bicarbonate co-transporter, in which case, the expression for the intracellular concentration will be the following:

$$n^i_{HCO_3} = n^o_{HCO_3} \exp\left(\varphi + \frac{\Delta\mu_A}{9}\right). \tag{2.18}$$

Subsequently, the dependence of the intracellular concentration of chlorine anions for the neurons of squid can be written as follows:

$$n^{in}_{Cl} = n^{out}_{Cl} \cdot \frac{n^{in}_{HCO}}{n^{out}_{HCO}} = n^{out}_{Cl} \exp\left(\varphi + \frac{\Delta\mu_A}{9}\right) \tag{2.19}$$

Two systems exist in neurons for the transport of calcium: the Ca^{2+}-ATPase and the Na^+-Ca^{2+}-exchanger. Furthermore, the cell membrane is permeable to these ions; therefore, passive penetration of ions occurs. The expressions for the fluxes will be similar to those that were recorded for cardiomyocytes, and the calculated values for the internal concentrations of this ion will be the following:

- for the neurons of squid:
  - Ca-ATP: $n^i_{Ca} = 3.7 \cdot 10^{-6}$ mM;
  - Na–Ca: $n^i_{Ca} = 3.2 \cdot 10^{-3}$ mM;
  - Passive flux: $n^i_{Ca} = 1.8 \cdot 10^3$ mM.

- for the neurons of mammals:
  - Ca-ATP: $n^i_{Ca} = 9.3 \cdot 10^{-7}$ mM;
  - Na–Ca: $n^i_{Ca} = 1.9 \cdot 10^{-4}$ mM;
  - Passive flux: $n^i_{Ca} = 453.2$ mM.

The minimum deviation from the experimental data (for both types of neurons, this value is equal to 0.0001 mM) is provided by the calcium-sodium exchanger; therefore, the expression for calcium ions is the following:

$$n^i_{Ca} = n^o_{Ca} \cdot \left(\frac{n^i_{Na}}{n^o_{Na}}\right)^3 \cdot \exp(\varphi) = n^o_{Ca} \cdot \exp(-2\varphi - \Delta\mu_A). \tag{2.20}$$

The resulting expression is identical to that of the dependence of the internal on external parameters for Ca-ATPase.

No information exists in the literature about the systems of active transport for magnesium in neurons. However, such a system should exist. If it is assumed that magnesium is transported in neurons only by a passive mechanism due to the doubly charged ion of Mg^{2+}, a very large concentration of magnesium ions within the cell will occur. Equation (2.9) can be used to determine the distribution of the passive transfer of magnesium:

$$n^i_{Mg} = n^o_{Mg} \exp(-2\varphi). \tag{2.21}$$

By substituting the values of the experimental data for the values of the potential and extracellular concentration of magnesium ions, we obtain the values

2.3 Neurons

of the internal concentration of magnesium ions: 9.1×10^3 mM for the squid neuron and 181.2 mM for mammalian neurons. This value indicates a clear discrepancy between the experimental data, which is 10 and 0.7 mM, respectively. A similar situation was considered previously for cardiac muscle cells (Melkikh and Sutormina 2008), and it was shown that passive transport of magnesium ions into the cell results in a high concentration of these ions inside the cell and a considerable reduction in the potential across the membrane. In heart muscle cells, a Na^+–Mg^{2+} exchanger is the primary method of transport of magnesium. In view of the above data and based on the considerable similarity of the transport systems in cardiac muscle cells and neurons, we assume that the same exchanger exists in neurons. The ion concentrations that are produced by this exchanger can be calculated by substituting the known experimental data for the neurons in (2.8):

$$n^i_{Mg} = n^o_{Mg} \exp(-\varphi) \frac{n^i_{Na}}{n^i_{Na}} = 109.3 \text{ mM for the neurons of squid, and}$$

$$n^i_{Mg} = n^o_{Mg} \exp(-\varphi) \frac{n^o_{Na}}{n^o_{Na}} = 1.3 \text{ mM for the neurons of mammals,}$$

which is in qualitative agreement with the experimental data. Thus, we can assume that the Na^+–Mg^{2+} exchanger is essential for the transport of magnesium. When the expression for the concentration of sodium ions is taken into account, the dependence on magnesium ions will take the form of (2.10).

By choosing the main transport system, we can derive an equation for the membrane potential of a neuron, which will depend only on the external concentrations of the main potential-forming ions. Note that for both types of neurons, we chose the same main transport systems except for chlorine anions. By substituting the expressions that are obtained for the ions, the condition of electroneutrality for the squid neuron can be written as follows:

$$n^o_{Ca} \cdot \exp(-2\varphi - \Delta\mu_A) + n^o_{Mg} \cdot \exp\left(-2\varphi - \frac{\Delta\mu_A}{3}\right) + n^o_{Na} \exp\left(-\frac{\Delta\mu_A}{3} - \varphi\right)$$
$$+ n^o_K \exp(-\varphi) = (n^o_{Cl} + n^o_{HCO3}) \exp\left(\varphi + \frac{\Delta\mu_A}{9}\right) + Z_A n_A;$$
(2.22)

and the condition for mammalian neurons can be written as follows:

$$n^o_{Ca} \cdot \exp(-2\varphi - \Delta\mu_A) + n^o_{Mg} \cdot \exp\left(-2\varphi - \frac{\Delta\mu_A}{3}\right) + n^o_{Na} \exp\left(-\frac{\Delta\mu_A}{3} - \varphi\right)$$
$$+ n^o_K \exp(-\varphi) = \left(n^o_{Cl} + n^o_{HCO3} \cdot \exp\left(\frac{\Delta\mu_A}{9}\right)\right) \exp(\varphi) + Z_A n_A.$$
(2.23)

When neglecting the divalent cations because of their small effect on the membrane potential, solutions of these equations can be written as the following:

- for the neurons of squid:

$$\varphi = -\ln\left(\sqrt{\left(\frac{Z_A n_A}{2\left(n^o_{Na}\cdot\exp\left(-\frac{\Delta\mu_A}{3}\right)+n^o_K\right)}\right)^2 + \frac{\left(n^o_{Cl}+n^o_{HCO_3}\right)\cdot\exp\left(\frac{\Delta\mu_A}{9}\right)}{2\left(n^o_{Na}\cdot\exp\left(-\frac{\Delta\mu_A}{3}\right)+n^o_K\right)} + \frac{Z_A n_A}{2\left(n^o_{Na}\cdot\exp\left(-\frac{\Delta\mu_A}{3}\right)+n^o_K\right)}}\right).$$

(2.24)

- for mammalian neurons:

$$\varphi = -\ln\left(\sqrt{\left(\frac{Z_A n_A}{2\left(n^o_{Na}\cdot\exp\left(-\frac{\Delta\mu_A}{3}\right)+n^o_K\right)}\right)^2 + \frac{n^o_{Cl}+n^o_{HCO_3}\cdot\exp\left(\frac{\Delta\mu_A}{9}\right)}{2\left(n^o_{Na}\cdot\exp\left(-\frac{\Delta\mu_A}{3}\right)+n^o_K\right)} + \frac{Z_A n_A}{2\left(n^o_{Na}\cdot\exp\left(-\frac{\Delta\mu_A}{3}\right)+n^o_K\right)}}\right).$$

(2.25)

Currently, the intracellular concentration of non-penetrative anions is unknown, but we can calculate this value by using the electro-neutrality condition of the internal environment:

$$n_A = n^i_{Na} + n^i_K - n^i_{Cl} - n^i_{HCO_3}.$$

When substituting the concentrations of sodium, potassium, bicarbonate and chloride ions outside of cells and the intracellular concentrations of non-penetrative ions, we obtain a value for the resting potential of a mammalian neuron of approximately—86.7 mV, and the potential value for the squid neuron would be approximately—85.1 mV. The obtained results from the calculations for the concentrations and potentials for squid and mammalian neurons are shown in Tables 2.5 and 2.6, respectively.

Although deviations from the experimental data are high, we have achieved satisfactory agreement between theory and experiment.

Table 2.5 Comparison of the experimental and calculated values of the internal ion concentrations and potential for squid neurons

Ions	Experimental data, mM	Calculated data, mM
Potassium	360	135
Sodium	69	7.3
Chlorine	157	158.7
Calcium	10^{-4}	3.7×10^{-6}
Magnesium	10	11.5
Bicarbonate	8	17.1
Potential, mV	$-(65 \div 70)$	-86.7

2.3 Neurons

Table 2.6 Comparison of experimental and calculated values for the internal ion concentrations and potentials of mammalian neurons

Ions	Experimental data, mM	Calculated data, mM
Potassium	150	74
Sodium	15	2.6
Chlorine	9	9,3
Calcium	10^{-4}	9.3×10^{-7}
Magnesium	0.7	0.2
Bicarbonate	8	17.1
Potential, mV	$-(65 \div 70)$	-85.1

2.3.2 Model of Ion Transport with a Restriction of Deviation from the Experimental Data

In some cases, the accuracy of calculating the internal concentrations of ions and resting potential based on the algorithm "one ion—one transport system" may not be sufficient. In this case, we use a special method to find dependencies of the intracellular on the external concentrations; and this method is referred to as "equivalent transporter". In a situation where it is not possible to choose a main mechanism of ion transport, we assume that two transporters work simultaneously and that when both work together, they give the most significant contribution to the overall result. In other words, two or more transport systems are replaced by one whose action is equivalent to their combined actions, as is schematically shown in Fig. 2.7:

The flux that is produced by an "equivalent mechanism" for the *j*th type of ion will be the following:

$$J = C_j \left(n^i \exp(\Delta \mu_j - A\varphi) - n^o \right) \tag{2.26}$$

where A is the charge of the transported ions, and $\Delta\mu_j$ is the EMF (electromotive force) of the equivalent mechanism. To find the parameter $\Delta\mu_j$, we can use the concentrations inside and outside of the cell that were obtained experimentally. The desired relationship is used to find the potential as a function of the extracellular concentration.

One type of ion, such as potassium ions, being simultaneously transported by several systems is much more difficult to solve analytically. In this case, unknown

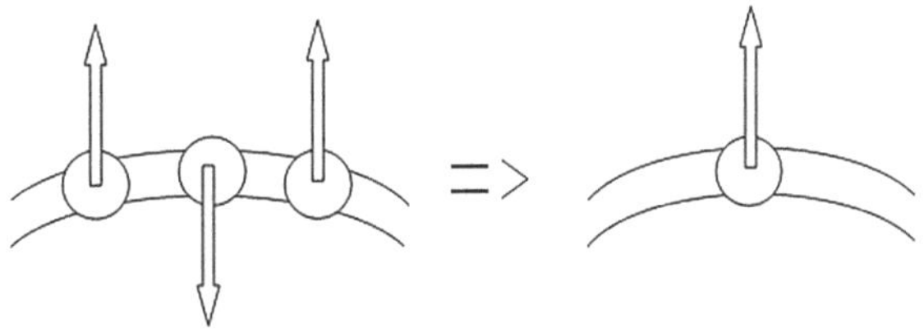

Fig. 2.7 Equivalent transport mechanism

constants, which take into account the speed of the pumps, appear in the equations. We make the following assumption, which can significantly improve the accuracy of the algorithm: combine several active transport mechanisms for potassium ions in an "equivalent pump." The result of such a pump would be the distribution of potassium ions inside the cell, which would differ from its passive distribution. The expression for the intracellular concentration, which takes into account active and passive transport, can be written as follows:

$$n_K^i = n_K^o \exp(\Delta\mu_K - \varphi) \qquad (2.27)$$

where $\Delta\mu_K$—is the effective potassium EMF. From this expression and by using known ion concentrations inside and outside the cell, we can find the value of the EMF of potassium ions and further to use this expression to calculate the potential and the intracellular concentrations of other ions. For the squid neuron, $\Delta\mu_K = 0.984$, and for mammalian neurons, $\Delta\mu_K = 0.706$.

The expression for calculating the intracellular concentration of sodium ions, which is generated by the Na–K-ATPase, takes into account the equivalent transport system for potassium ions and takes the following form:

$$n_{Na}^{in} = n_{Na}^{out} \exp\left(\frac{2 \cdot \Delta\mu_K}{3} - \varphi - \frac{\Delta\mu_A}{3}\right). \qquad (2.28)$$

By substituting the known values and the calculated EMF, we obtain the following values for the intracellular concentrations of sodium ions: 14 mM for the neurons of squid and 4.1 mM for mammalian neurons. The obtained values correspond better to the experimental data (69 and 15 mM, respectively) than those that were calculated in the previous section.

The values of the concentrations of calcium ions, which were generated by the sodium-calcium exchanger, can be recalculated by altering the expression for sodium ions:

$$n_{Ca}^i = \exp(\varphi) \cdot n_{Ca}^o \cdot \left(\frac{n_{Na}^i}{n_{Na}^o}\right)^3 = n_{Ca}^o \exp(-2 \cdot \varphi - \Delta\mu_A + 2 \cdot \Delta\mu_K) \qquad (2.29)$$

By substituting the known data, it is evident that the intracellular concentration of calcium ions in the neurons of squid will be equal to 2.7×10^{-5} mM, and in mammalian neurons, this concentration would be equal to 3.8×10^{-6} mM (0.0001 mM is the experimental value in both neurons).

Similarly, the dependence of intracellular on external concentrations of magnesium ions will change as follows:

$$n_{Mg}^i = n_{Mg}^o \cdot \exp(-\varphi) \cdot \frac{n_{Na}^i}{n_{Na}^o} = n_{Mg}^o \cdot \exp\left(\frac{2}{3} \cdot \Delta\mu_K - 2\varphi - \frac{\Delta\mu_A}{3}\right) \qquad (2.30)$$

As a result, for these ions, we find that in squid neurons, the intracellular concentration is equal to 22.2 mM, and in mammalian neurons, this value is 0.4 mM (the experimental data is 10 and 0.7 mM, respectively).

2.3 Neurons

The condition of electro-neutrality can be written as follows:

- for the neurons of squid:

$$n_{Na}^\circ \exp\left(\frac{2\cdot\Delta\mu_K}{3} - \frac{\Delta\mu_A}{3} - \varphi\right) + n_K^\circ \exp(\Delta\mu_K - \varphi)$$
$$= \left(n_{Cl}^\circ + n_{HCO_3}^\circ\right)\exp\left(\varphi + \frac{\Delta\mu_A}{9}\right) + Z_A n_A \quad (2.31)$$

- for the neurons of mammals:

$$n_{Na}^\circ \exp\left(\frac{2\cdot\Delta\mu_K}{3} - \frac{\Delta\mu_A}{3} - \varphi\right) + n_K^\circ \exp(\Delta\mu_K - \varphi)$$
$$= \left(n_{Cl}^\circ + n_{HCO3}^\circ \cdot \exp\left(\frac{\Delta\mu_A}{9}\right)\right)\cdot \exp(\varphi) + Z_A n_A \quad (2.32)$$

The explicit expressions for the potential will be written as follows:

- for the neurons of squid:

$$\varphi = -\ln\left(\sqrt{\left(\frac{Z_A n_A}{2\left(n_{Na}^\circ \cdot \exp\left(\frac{2\Delta\mu_K}{3} - \frac{\Delta\mu_A}{3}\right) + n_K^\circ \cdot \exp(\Delta\mu_K)\right)}\right)^2 + \frac{(n_{Cl}^\circ + n_{HCO3}^\circ)\cdot \exp(\Delta\mu_A/9)}{2\left(n_{Na}^\circ \cdot \exp\left(\frac{2\Delta\mu_K}{3} - \frac{\Delta\mu_A}{3}\right) + n_K^\circ \cdot \exp(\Delta\mu_K)\right)} + \frac{Z_A n_A}{2\left(n_{Na}^\circ \cdot \exp\left(\frac{2\Delta\mu_K}{3} - \frac{\Delta\mu_A}{3}\right) + n_K^\circ \cdot \exp(\Delta\mu_K)\right)}\right)$$

(2.33)

- for the neurons of mammals:

$$\varphi = -\ln\left(\sqrt{\left(\frac{Z_A n_A}{2\left(n_{Na}^\circ \cdot \exp\left(\frac{2\Delta\mu_K}{3} - \frac{\Delta\mu_A}{3}\right) + n_K^\circ \cdot \exp(\Delta\mu_K)\right)}\right)^2 + \frac{n_{Cl}^\circ + n_{HCO_3}^\circ \cdot \exp(\Delta\mu_A/9)}{2\left(n_{Na}^\circ \cdot \exp\left(\frac{2\Delta\mu_K}{3} - \frac{\Delta\mu_A}{3}\right) + n_K^\circ \cdot \exp(\Delta\mu_K)\right)} + \frac{Z_A n_A}{2\left(n_{Na}^\circ \cdot \exp\left(\frac{2\Delta\mu_K}{3} - \frac{\Delta\mu_A}{3}\right) + n_K^\circ \cdot \exp(\Delta\mu_K)\right)}\right)$$

(2.34)

By substituting the external concentrations of the main ions and the calculated EMF for potassium, we obtain the following values of the potential: -68.7 mV for

Table 2.7 Comparison of the experimental and calculated values of the internal ion concentrations and potential for squid neurons

Ions	Experimental data, mM	Calculated data, mM
Sodium	69	14
Calcium	10^{-4}	2.7×10^{-5}
Magnesium	10	22.2
Potential, mV	$-(65 \div 70)$	-68.7

Table 2.8 Comparison of the experimental and calculated values for the internal ion concentrations and potential for mammalian neurons

Ions	Experimental data, mM	Calculated data, mM
Potassium	15	4.1
Calcium	10^{-4}	3.8×10^{-6}
Magnesium	0.7	0.4
Potential, mV	$-(65 \div 70)$	-70.7

squid neurons and -70.7 mV for mammalian neurons. These results are in good agreement with the experimental data.

The results from calculating the concentrations and potential when using an "equivalent potassium pump" for squid and mammalian neurons are shown in Tables 2.7 and 2.8, respectively. The tables show only those ions with altered concentrations by using the algorithm for the potassium EMF.

2.3.3 Regulation of Ion Transport

According to our algorithm, we can construct a model for the regulation of ion transport in the neurons of mammals. Transport systems in neurons, such as the sodium ion transport system, were identified earlier, and graphs of this ion transport mechanism are shown on Fig. 2.8.

The figure shows the intracellular concentration after change α of sodium ions in the external medium. Symbols in the figure are as follows: the solid red line indicates the required value of the intracellular concentration, blue dots represent the work of the K–Na-ATPase, the green dotted line represents the work of the co-transporter for Na–HCO$_3$; and the purple dash-dot line refers to the Na–Ca-exchanger.

When switching transport systems, it is advantageous for a cell to maintain only one active mechanism to be effective in energy consumption. In this case, a strategy of switching must be constructed that gives the smallest deviation from the desired value, assuming that one transport system works (Fig. 2.9).

The graph shows that when the external sodium ion concentration is reduced by about approximately 30 % of the normal level the calcium ion exchanger works.

Fig. 2.8 Mechanisms of sodium transfer in neurons

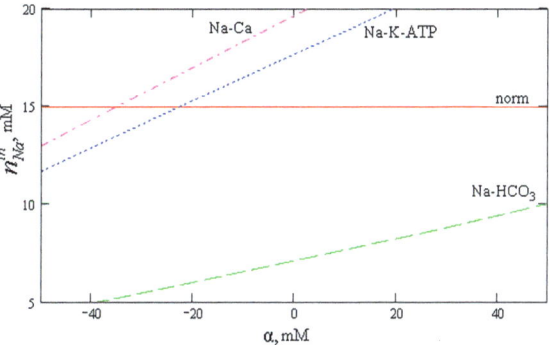

Fig. 2.9 Strategy of sodium ion transport in neuron regulation

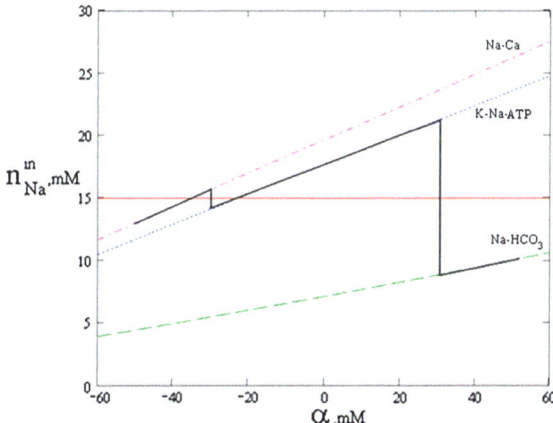

Furthermore, the ATP-dependent pump effectively works in a sufficiently large range of extracellular concentrations of sodium but only when in excess of the normal value. When the external sodium concentration is large, the Na–HCO$_3$ symporter works.

However, during such a strategy, a significant change in the intracellular concentration of sodium takes place, which results in a deviation from the normal value by nearly 50 % and may be detrimental to vital cells. If the tolerance of the internal concentration of sodium is limited to an interval of 30 % in both directions, then it needs to perform ion transport by two mechanisms in a range of external concentrations (Fig. 2.10).

In the figure, the dotted line separates the region of tolerance values for the intracellular concentrations of sodium. It is observed when the external sodium ion concentration increases by 20 mM or more to maintain the desired value of the internal concentration is only possible by the simultaneous operation of the ATP-dependent pump and Na–HCO$_3$ co-transporter. Additionally, when this increase (more than 55 mM $\approx$ 40 % of the normal value) exceeds the external concentration of sodium ions, the sodium-bicarbonate co-transporter can maintain the necessary internal concentration in the absence of other transport systems.

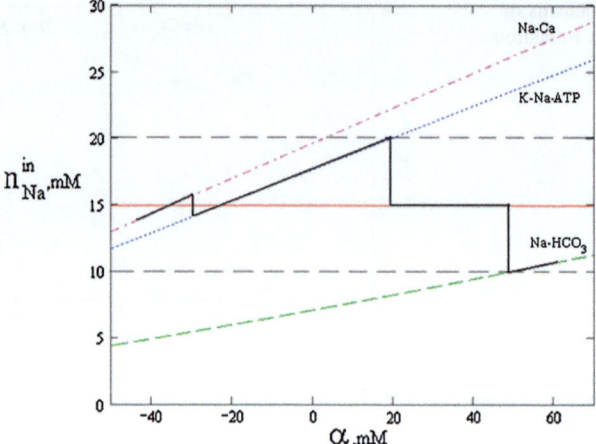

Fig. 2.10 Strategy of sodium ion regulation with a restriction of maximum deviations from the normal values

Graphs show how transport systems work at the changes in the extracellular concentrations (Fig. 2.11).

From the figure and by taking into account (2.17), it is evident that in a certain range of external low concentrations of sodium, the Na^+–Ca^{2+}-exchanger is sufficient. When the ion content in the external environment increases, the ATP-dependent pump requires for maintaining ion balance; co-transport is able to maintain the necessary intracellular concentration only after a greater increase in the extracellular concentration of sodium ions.

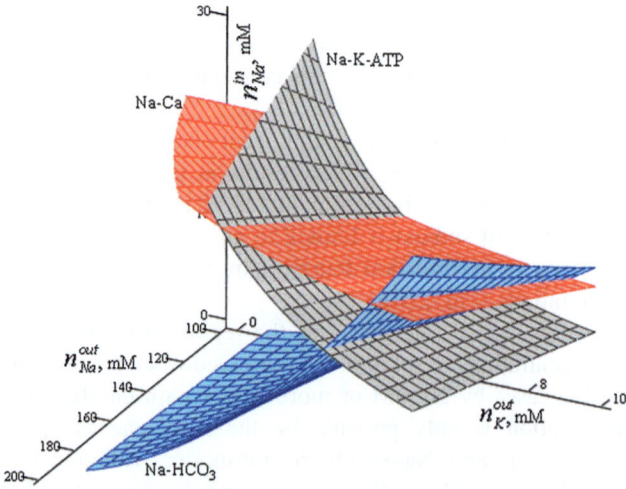

Fig. 2.11 Systems of sodium transport in neurons as a function of the external concentrations of two ions

2.4 Erythrocytes

The current red blood cell model includes (see, for example, (Jamshidi et al. 2001)) 36 independent variables that are described in the main biochemical pathways in the human erythrocyte. Transport of ions through the red blood cell (RBC) membrane is also included in this system of equations (Werner and Heinrich 1985; Joshi and Palsson 1989; Jamshidi et al. 2002; Mulquiney et al. 1999; Lew and Bookchin 1986). The goal of systems biology is a comprehensive description of a cell (or an organism) through mathematical methods by the use of computers. Currently, systems biology of RBCs is concerned mainly with gene and metabolic networks of cells; however, the transport subsystem of RBCs has not been studied sufficiently.

2.4.1 Model of Ion Transport

In red blood cells, ion transport includes the following transport systems (Freedman 2001; El-Mallakh 2004; Freedman and Hoffman 1979): ATP-dependent pumps (Ca^{2+}-ATP-pump; Na^+-K^+-ATP-pump), exchangers (K^+–Cl^-–symporter; Na^+–Ca^{2+}—antiporter; K^+–Na^+–Cl^-—symporter; Cl^-–Zn^{2+}—symporter; HCO_3^- Cl^-—antiporter; and Na^+–Li^+—antiporter), and passive transportation for all types of ions (Fig. 2.12).

Based on the proposed algorithm, we shall discuss the transport systems that transfer potential-generating ions in a red blood cell. Because the nonequilibrium state in a cell is provided by the consumption of energy in the form of ATP, it is necessary to choose an ATP-dependent pump as one of the main methods of transport. Therefore, the expression for the flux of sodium ions that is created by the pump may be written as the following equation:

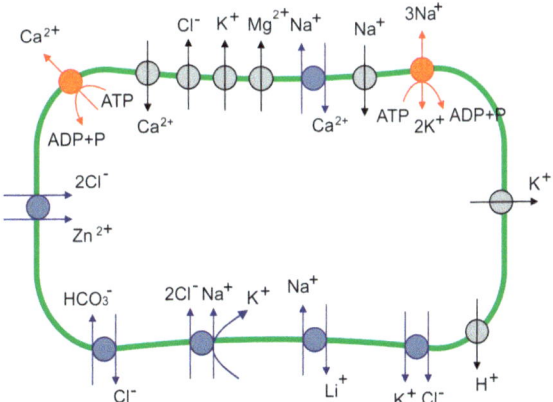

Fig. 2.12 Ion transport systems in RBCs

$$J_{Na-K-ATP} = C_{Na-K-ATP}\left[\exp(\Delta\mu_A + \varphi)(n_{Na}^i)^3(n_K^o)^2 - (n_{Na}^o)^3(n_K^i)^2\right], \quad (2.35)$$

where $C_{Na-K-ATP}$ is the constant of the active transport of sodium and potassium ions.

This system may be the main method of transport for sodium or potassium ions. By setting the flux (2.35) equal to zero and using the experimental data on the potassium ions inside and outside of a cell and the extracellular sodium ion concentration, we calculate the concentration of internal sodium to be $n_{Na}^i = 6.1$ mM. Similarly, we calculate the concentration of potassium ions, which is created by this transport system when in equilibrium, as $n_K^i = 279$ mM. After comparing both calculated quantities with the experimental data ($n_{Na}^i = 10$ mM, $n_K^i = 135$ mM), we chose the ATP-dependent pump as the main method of transport of sodium ions.

Three systems exist that transport potassium ions: Na⁺–K⁺-ATPase, and K⁺–Cl⁻ and K⁺–Na⁺–Cl⁻—co-transporters, and the permeability of the membrane provides a passive flux of ions along the electrochemical gradient. The expressions for the fluxes that are created by other mechanisms (except ATPase) can be written separately.

$$J_{K-Cl} = C_{K-Cl} \cdot \left[n_K^i \cdot n_{Cl}^i - n_K^o \cdot n_{Cl}^o\right], \quad (2.36)$$

$$J_{K-Na-Cl} = C_{K-Na-Cl} \cdot \left[n_{Na}^i \cdot n_K^i \cdot (n_{Cl}^i)^2 - n_{Na}^o \cdot n_K^o \cdot (n_{Cl}^o)^2\right], \quad (2.37)$$

$$J_{Kpas} = P_K \cdot \left[n_K^i \cdot \exp(\varphi) - n_K^o\right], \quad (2.38)$$

here and elsewhere, P_i is the constant of passive transport for the ith type of ions along their electrochemical gradient. The values of the internal concentrations of potassium ions that are created by each mechanism are as follows:

- K–Cl: $n_K^i = 5.1$ mM;
- K–Na–Cl: $n_K^i = 97.5$ mM;
- passive flow: $n_K^i = 5.7$ mM.

After comparing the results with the experimental data (135 mM), we determine that the K⁺–Na⁺–Cl⁻—co-transporter is the main transport system for potassium. We can obtain the explicit form of the dependencies of the internal on the external concentrations and the membrane potential for Na⁺ and K⁺ by solving the system of two equations that describe the work of the mechanisms that were chosen: the ATPase and co-transporter. However, doing so requires the explicit form of the dependence of the internal on the external parameters for chlorine ions.

Similarly, we shall consider chlorine ion transport, which is provided by the following co-transporters: K⁺–Cl⁻, K⁺–Na⁺–Cl⁻, and Cl⁻–Zn²⁺, the anti-porter $HCO_3^-Cl^-$ and passive flux through a membrane. By comparing those types of transport with the experimental data, we can conclude that the largest contribution to the required concentration of chlorine ions is made by its passive flux or their

2.4 Erythrocytes

exchange for bicarbonate ions. In this case, chlorine ions are transported passively, and the dependence of the internal on the external concentration will be represented by the following equation:

$$n^i_{Cl} = n^o_{Cl} \exp(\varphi). \qquad (2.39)$$

The explicit forms of the dependencies of the internal concentrations of sodium and potassium ions can be written as the following equations:

$$n^i_{Na} = n^o_{Na} \exp\left(-\frac{3}{5}\varphi - \frac{\Delta\mu_A}{5}\right), \qquad (2.40)$$

$$n^i_K = n^o_K \exp\left(-\frac{2}{5}\varphi + \frac{\Delta\mu_A}{5}\right). \qquad (2.41)$$

The transport of bicarbonate ions across the red blood cell membrane, apart from passive penetration, is implemented by their exchange with chlorine ions. Calculations showed that the required concentration is created by passive flux, and the expression for the internal concentration of HCO_3 will be the following:

$$n^i_{HCO_3} = n^o_{HCO_3} \exp(\varphi). \qquad (2.42)$$

There is currently no published information concerning the systems of active magnesium ion transport. By considering the transport of these ions as passive, we determine the value of the internal magnesium ion concentration to be 3.5 mM. These results satisfactorily coincide with the experimental data (2.5 mM) for magnesium ions; therefore, we may consider passive flux as the main transport system for this type of ion.

The transport of lithium ions into a cell is provided by its exchange with sodium; however, passive flux of these ions also exists. After comparing the internal concentration of lithium ions (0.7 ÷ 1.0 mM) that is created separately (0.007 mM and 0.06 mM, respectively), we determined that the passive flow of these ions along their electrochemical gradient is the main transport system for such ions.

Similar reasoning for protons gives the value of their internal concentration as 0.061 mM, and these results satisfactorily coincide with the experimental data ($n^{in}_H = 0.062$ mM). Therefore, we may consider passive flux as the main transport system for protons.

The flux of zinc ions is provided by the work of a co-transporter that also transports chlorine and passive membrane permeability. However, the calculated value of the internal concentration that is created by the co-transporter better coincides with the experiment results.

Three ways to transport doubly charged calcium cations in a red blood cell exist: active transport with ATP energy consumption, passive transport, and transport in exchange for sodium ions.

- Ca-ATP: $n^i_{Ca} = 9.8 \cdot 10^{-8}$ mM;
- Na–Ca: $n^i_{Ca} = 2.4 \cdot 10^{-4}$ mM;
- passive flow: $n^i_{Ca} = 2.4$ mM.

The best coincidence with the experimental data (i.e., the Ca^{2+} concentration in a red blood cell ranges from $(30 \div 60) \times 10^{-6}$ mM) is provided by the Na$^+$–Ca^{2+} exchanger. Thus, the expression for the dependence of the internal concentration of sodium ions on the membrane potential and external concentration of sodium ions will be the following:

$$n^i_{Ca} = \exp(\varphi) \cdot n^o_{Ca} \left(\frac{n^i_{Na}}{n^o_{Na}}\right)^3 = n^o_{Ca} \cdot \exp\left(-\frac{4}{5} \cdot \varphi - \frac{3 \cdot \Delta \mu_A}{5}\right) \quad (2.43)$$

Thus, we can calculate the intracellular concentrations of the main ions that are transported through the RBC membrane (Freedman 2001; El-Mallakh 2004; Freedman and Hoffman 1979), and these obtained values have been tabulated (Table 2.9).

When taking into account all the ion types that are present in a cell, the electroneutrality condition for the internal environment of a cell can be written as the following equation:

$$n^{in}_{Na} + n^{in}_{K} + 2 n^{in}_{Ca} + 2 n^{in}_{Mg} - n^{in}_{Cl} - n^{in}_{HCO} + n^{in}_{Li} + n^{in}_{H} + 2 n^{in}_{Zn} - Z_A n_A = 0, \quad (2.44)$$

where $Z_A n_A$ is the product of the charge and the concentration of non-penetrating, intracellular anions. From this expression, we can obtain the analytical dependency of the resting potential of the cell membrane as a function of the external concentration; however, the conclusion for analytical dependency seems to be complex if we do not neglect bivalent cations.

After numerically solving the expression, we calculate the potential of the red blood cell membrane to be -10.88 mV (Freedman 2001), which is in good agreement with the experimental data.

Table 2.9 Comparison of the experimental and calculated values of the internal concentrations of ions in RBCs

Ions	Experimental data, mM	Calculated* data, mM
Sodium	10	6.1
Potassium	135	97.5
Chlorine	78	69.6
Bicarbonate	16	16.3
Magnesium	2.5	3.5
Lithium	$0.7 \div 1.0$	0.06
Protons	0.062	0.061
Zink	0.024	0.028
Calcium	$(30 \div 60) \times 10^{-6}$	53×10^{-6}

*The membrane potential value (-11 mV) that was obtained experimentally was used

2.4 Erythrocytes

The calculations were made using $\Delta\mu_A = 17$, which is based on the best agreement of the concentrations and potential with the experimental data; however, this value is somewhat less than the conventional value of $\Delta\mu_A = 20$ for different types of cells. It is necessary to take into consideration that RBCs do not have mitochondria; therefore, ATP molecules should be transported inside this organelle by specific proteins. Hence, we can conclude that in a red blood cells, the $\Delta\mu_A$ should be less than in other cells.

2.4.2 Model of Regulation of Ion Transport: Efficiency or Robustness?

To simulate the processes of regulation, we shall consider the two extremities and realize that the actual cell behavior strategies are a combination of these behaviors.

The model that was constructed in Sect. 2.3.1 largely coincides with the experimental data; however, we cannot affirm that when environmental changes occur, this model will give reliable results.

According to the proposed algorithm, all of the transport systems for potassium ions in a red blood cell are considered, and the internal concentrations of potassium ions form dependencies on the external ions for each transport system. In doing so, we shall also take into consideration the potential change (Fig. 2.13).

The figure shows the dependencies of the internal concentration of potassium ions on an α change of K ion concentration in the environment. The figure legend is as follows: the required value of the intracellular concentration is shown by a solid red line; the work of the K–Na-ATPase is represented by blue dots; and K–Na–Cl-co-transport is represented by a green, dashed line.

For energy-effective work, it is profitable for a cell to keep only one transport mechanism functional. A strategy of switching for this case supposes that an ion transport system works that gives the least deviation from the required value (Fig. 2.14).

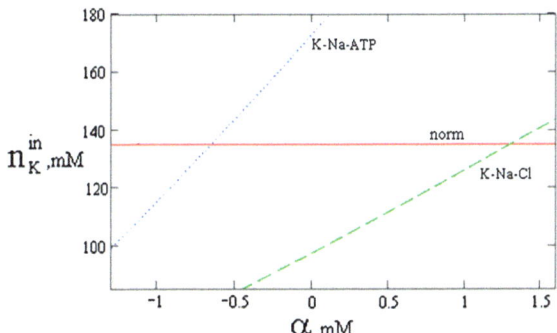

Fig. 2.13 The dependencies of the internal concentration of potassium ions on K–Cl α, which are additive in the external environment when two different transport systems are functional

Fig. 2.14 The strategy of regulation of potassium ion transport. α is the change of the external concentration of potassium ions

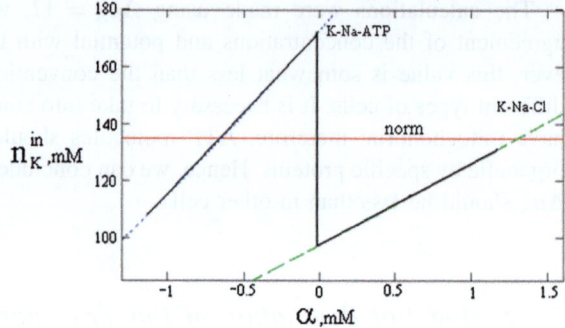

The dotted and dashed lines show the changes in the internal concentration that are conditioned by the work of the K–Na-ATPase and K–Na-Cl-co-transporters.

The graph shows that when the extracellular concentration of potassium ions is reduced, the ATP dependent pump works, but when this concentration increases, the K–Na–Cl-co-transporter is switched on. We make this conclusion because, at a deviation of 35 %, only one transport system is required to maintain the internal concentration.

Generally, a concentration change of at least one ion outside of a cell results in a change in the concentrations of all of the internal ions due to a change in potential. Therefore, it is important to study the dependence of the concentration of one ion in a cell on the change in the concentrations of two ions outside of the cell. The dependence of the intracellular concentration of potassium ions on the change of the external concentrations of potassium and sodium ions was plotted. In this case, the sought-for strategy will be the one that is maximally effective at each moment when only one transport mechanism is functional and will consist of a set of dots on planes that designate the mechanisms of ion transport that differ minimally from the normal value in relative units. Thus, for each value of extracellular concentration of an ion, a point on one plane will represent the work of the transport mechanisms, which is as close as possible to the specified parameters.

The algorithm for simulation of the ion transport regulation processes in a cell for a three-dimensional case is similar to that for a two-dimensional case.

From the Fig. 2.15, we can conclude that when the concentration of potassium ions is low in the extracellular environment, it is necessary to spend energy to provide the required intracellular concentration for any amount of sodium ions. In other words, the difference in Na^+ concentrations is not sufficient to maintain the normal level of potassium ions in a cell. When K^+ levels are sufficient in the cellular environment, a switch takes place from an ATPase-dependent system to a co-transporter-dependent system, and the energy consumption for maintaining the intracellular balance of K^+ concentrations decreases, which indirectly confirms our supposition concerning "selection" by a cell for the most effective systems of ion transport. Furthermore, the co-transporter switches on when the external concentration of potassium ions increases.

2.4 Erythrocytes

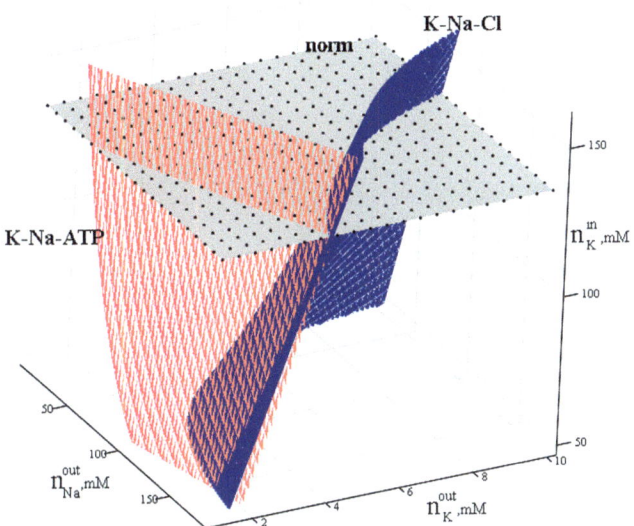

Fig. 2.15 The dependence of the internal concentrations of potassium and sodium ions in the environment for different transport mechanisms

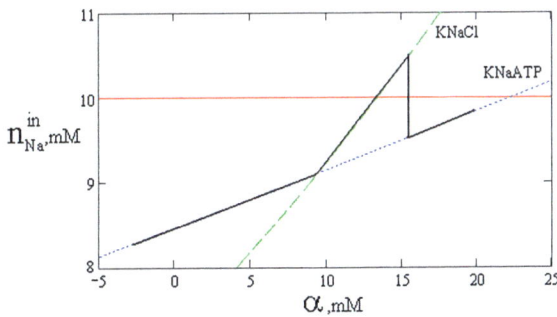

Fig. 2.16 The strategies for Na$^+$ transport regulation

Similarly, the transportation of Na$^+$ will be discussed with the supposition that sodium is added with chlorine. Two systems for Na$^+$ transport are capable of creating concentrations of this ion that are similar to the experimental value: the K–Na-ATPases and the K–Na–Cl-co-transporter.

In Fig. 2.16 α is the change in the Na$^+$ external concentration; and the dotted and dashed lines indicate the changes in the internal concentration that are conditioned by the work of the K–Na-ATPases and K–Na–Cl-co-transporter, respectively.

As follows from Fig. 2.16, the use of the K–Na-ATPase is the optimal mechanism for a defined concentration, but the K–Na–Cl-co-transporter functions when the external concentration of sodium ions increases to above 10 % of the normal value. Afterward, the ATP dependent pump switches on again.

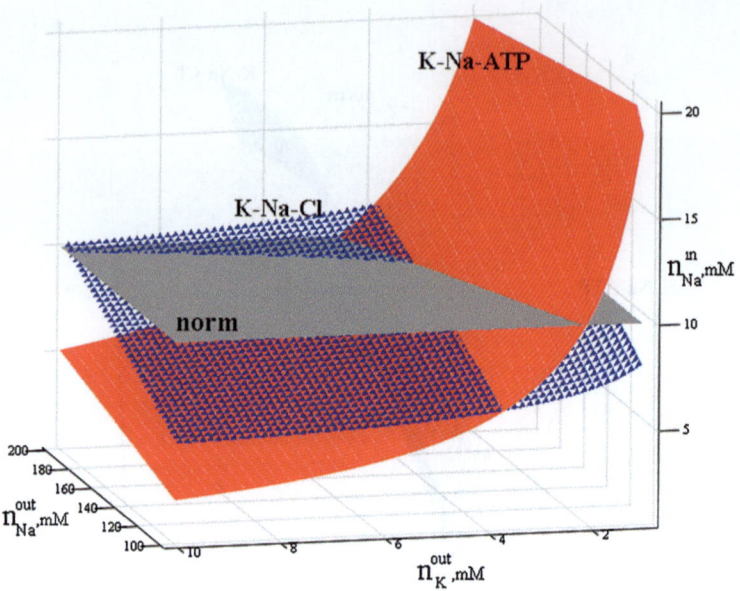

Fig. 2.17 The Na$^+$ transport systems as functions of the extracellular concentrations of two ions

The three-dimensional model in the figure shows that the main transport system for sodium ions is the ATP-dependent pump. In the short range of small values of K$^+$ external concentrations and when the external concentrations of both cations increase considerably (by approximately 30 % or more), K–Na–Cl-co-transport is sufficient to provide the required concentration of external Na$^+$ ions.

As is clear from Figs. 2.15 and 2.17, when there is an excessive amount of both cations in the environment, their homeostasis in a cell can be provided without energy consumption. But when the values of their external concentrations are small, ATP energy is required.

According to the experimental data, the speed of the different transport systems (including red blood cells) depends significantly on the ionic composition of their cells and their environment. For example, according to previous work (Freedman 2001, Lluch et al. 1996, Sriboonlue et al. 2005), the rate of the Na$^+$–K$^+$ -pump and Na$^+$–K$^+$–Cl$^-$-cotransporter in erythrocytes significantly changes when the environmental NaCl concentration is altered, i.e., switching from one transport system to another takes place. In addition, the cell volume depends on the concentration of NaCl or KCl in the environment. For example, regulation of the volume that occurs due to an increase (or decrease) in the speed of the various transport systems is well known for a variety of cells (see Alvarez-Leefmans 2001; Garcia-Romeu et al. 1991; Hoffman and Dunham 1995; Swietach et al. 2010; MacManus et al. 1995; Yachie-Kinoshita et al. 2010) and in particular, for red blood cells (Freedman 2001, Baumgarten and Feher 2001; Sarkadi and Parker 1991; Gusev et al. 1996).

2.5 Hepatocytes

Fig. 2.18 Mechanisms of ion transport in hepatocytes

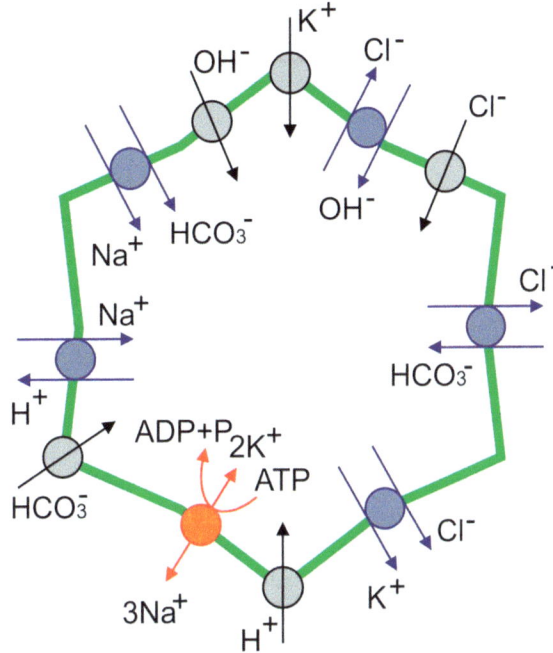

2.5 Hepatocytes

A model for active transport of ions across the membrane of hepatocytes has been published previously (Melkikh and Sutormina 2010). Here, we only present the main results that are needed to analyze the possible strategies for regulation.

Several transport systems exist in hepatocytes (Fossat et al. 1997; Murphy et al. 1980; Furimsky et al. 2000) such as the Na–K-ATP pump, the Na–H, Cl–HCO_3 and OH–Cl exchangers, and the Na–HCO_3 and K–Cl co-transporters. All of these systems are shown schematically in Fig. 2.18.

We will consider only the transport of ions that play a major role in the formation of the resting potential on the cell membrane. Furthermore, we will consider only K^+, Na^+, Cl^- and HCO_3^- as the main ions for this cell type. Other ions will be neglected because their contributions to the potential are negligible. In Table 2.10, the values of the concentrations of the major ions and the resting potential for rat liver cells according to Sillau et al. (1996) are listed.

2.5.1 Model for Ion Transport

Four transport systems are known to carry sodium ions in hepatocytes: the Na–K-ATPase, the Na–H exchanger, the Na–HCO_3 co-transporter and passive flow. The

Table 2.10 Experimental data for ion concentrations and potential in hepatocytes

Ions	Internal concentrations, mM	External concentrations, mM
Sodium	29	143
Potassium	166	4
Chlorine	20	120
Bicarbonate	25	27
Potential, mV		−49.8

contribution of the ATP-dependent pump in the formation of the internal sodium ion concentration will be considered as the most significant, and the pump itself is the main transport mechanism for this type of ion.

When selecting the main transport system for potassium ions, we assumed that hepatocytes act similar to neurons and introduced an "equivalent transporter", which is an expression that is similar to (2.28). Using the experimental value of the concentration of potassium ions on both sides of membrane we calculated that the value of the potassium EMF must be 1.73.

The dependence of the internal concentration of sodium ions on the external concentration, when an "equivalent transporter" is used, of potassium ions will be similar to that shown for the neuron (2.29).

The transport of chloride ions is provided by the Cl–HCO$_3$ and OH–Cl exchangers, the K–Cl co-transporter and passive transport through the membrane. The experimental data best fit the concentration that is created due to passive penetration of chloride ions through the membrane on the electrochemical gradient.

Transport of bicarbonate is provided by addition to the passive flow by the Cl–HCO$_3$ exchanger and Na–HCO$_3$ co-transporter, and the Cl–HCO$_3$ exchanger provided the minimum deviation from the experimental data.

For further calculations, the concentrations of non-penetrating anions must be determined. To calculate these concentrations, it is assumed that the pressure difference inside and outside of a mammalian cell can be neglected. Therefore, the condition of pressure equality on the membrane is written as follows:

$$n^i_{Na} + n^i_K + n^i_{Cl} + n^i_{HCO3} + n_A = n^o_{Na} + n^o_K + n^o_{Cl} + n^o_{HCO_3} \qquad (2.45)$$

From this equation, the concentration of non-penetrating ions can be calculated. Additionally, if the expressions that are obtained for the intracellular concentration of chloride ions, sodium, potassium, and bicarbonate are substituted, we obtain the following expression for the potential:

$$n^o_K \exp(\Delta\mu_K - \varphi) + n^o_{Na} \exp\left(2\frac{\Delta\mu_K}{3} - \varphi - \frac{\Delta\mu_A}{3}\right) \\ - n^o_{Cl} \cdot \exp(\varphi) - n^o_{HCO_3} \cdot \exp(\varphi) - Z_A n_A = 0 \qquad (2.46)$$

Then, the analytic dependence of the resting potential of cells on the external concentration is represented by the following equation:

2.5 Hepatocytes

Fig. 2.19 Changing of potential in hepatocytes is dependent on the addition of potassium ions into the environment

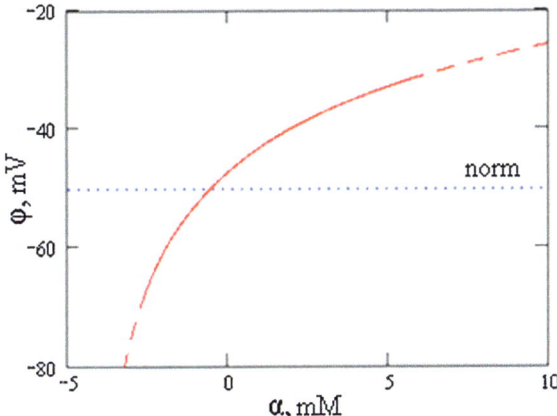

$$\varphi = \ln \left[\sqrt{\left(\frac{-Z_A n_A}{2 \cdot \left(n^o_{Cl} + n^o_{HCO_3}\right)} \right)^2 + \frac{n^o_K \cdot \exp(\Delta \mu_K) + n^o_{Na} \cdot \exp\left(\frac{2\Delta \mu_K}{3} - \frac{\Delta \mu_A}{3}\right)}{2 \cdot \left(n^o_{Cl} + n^o_{HCO_3}\right)}} - \frac{Z_A n_A}{2 \cdot \left(n^o_{Cl} + n^o_{HCO_3}\right)} \right] \quad (2.47)$$

By numerically solving this equation, we obtain the potential value of −47.4 mV, which agrees well with the experimental data (Fig. 2.19).

The Fig. 2.20 shows the dependence of the resting membrane potential (solid line) on the addition of α (in mM) into the environment by potassium ions with chlorine. The experimental value of the potential for hepatocyte cells is shown by dotted lines.

Table 2.11 summarizes the calculated values in comparison with the experimental data.

Table 2.11 clearly shows that the expressions that are obtained for the resting potential of the cells and those of the concentrations of chloride ions are in good agreement with the experimental data. However, a discrepancy exists between the calculated results and the experimental data for sodium and bicarbonate, which is possibly due to the lack of completeness of the information on systems of transport in hepatocytes.

Table 2.11 Comparison of experimental and calculated values for the internal ion concentrations and potential for hepatocytes

Ions	Experimental data, mM	Calculated data, mM
Sodium	29	4.3
Chlorine	20	16.2
Bicarbonate	25	4.5
Potential, mV	−49.8	−47.4

2.5.2 Regulation of Ion Transport

A model for the Regulation of Ion Transport in liver cells using all of the transport mechanisms that were introduced in the previous section can be constructed based on the model for sodium ions. Graphically, the work of each transport system when the external concentration of sodium ions change by the value of α (in mM) is shown in Fig. 2.20.

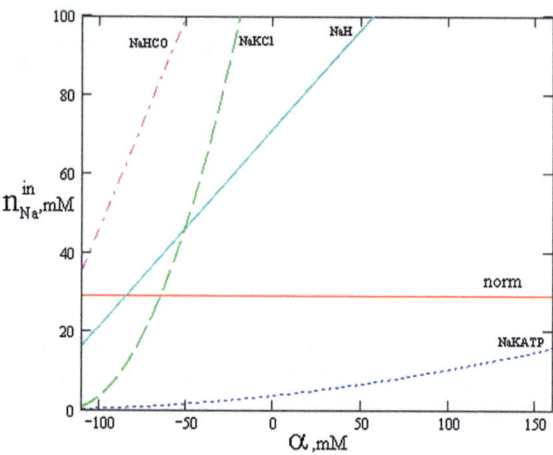

Fig. 2.20 Dependences of the internal concentrations of potassium on the external concentrations when different transport systems are functioning

The figure shows that none of the transport systems are in good agreement with the required value for the internal concentration of sodium ions. Additionally, in the previous section, calculations were carried out to support this finding. It can be concluded that to maintain the required concentration of intracellular sodium ions in hepatocytes, two transport mechanisms are required to function at the same time (Fig. 2.21).

Three-dimensional modeling confirms our conclusions on the basic functions of the ATP-dependent pump in forming the desired concentration of sodium ions and on the activation of the K–Na–Cl and Na–HCO$_3$ co-transporters with decreasing sodium content in the extracellular medium. However, the assumption that the Na–H-antiporter functions independently at low concentrations of sodium ions outside the cell is not confirmed. Therefore, we can conclude that the role of the antiporter is only regulatory, and this transporter switches only in addition to the co-transporter of sodium ions with bicarbonate. Importantly, experiments have shown that these transporters are activated together.

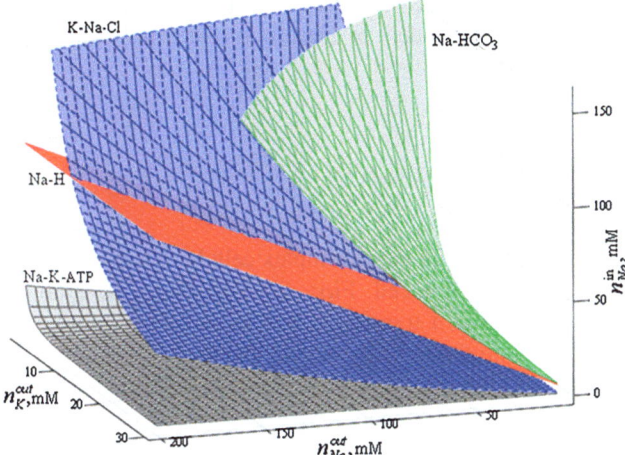

Fig. 2.21 Systems of transport for sodium ions as the function of the external concentrations of two ions

2.6 Regulation of Ion Transport in Compartments of a Mammalian Cell

Cells of living organisms contain a large number of compartments that are surrounded by membranes and are the basis of cell activity. In these compartments, the main processes for life function in the exchange of matter, energy and entropy between the organism and the environment. These compartments also divide the totality of the vital processes of cells into separate stages. The rough endoplasmic reticulum and ribosomes provide protein synthesis in accordance with the structure of RNA. In the Golgi apparatus, the conversion of tertiary and quaternary protein structures and the formation of markers that govern their directional transport occurs, and synaptic vesicles store neurotransmitters that are necessary for the transmission of nerve impulses. The mitochondria convert energy from the oxidation of food to the free energy of ATP, and ATP hydrolysis is a universal source of free energy at all stages of life.

Each of these stages of life, and many other processes, requires special conditions (concentrations of substances, osmotic pressure, electric potential) for their implementation.

In this regard, the maintenance of the concentrations of certain types of ions in cellular compartments plays an important role in the functioning of the cell as a whole. Transport systems have many different compartments that match the structure and properties of elements (specialized proteins), and the general ideology of constructing models of transport for substances in these compartments is described by Melkikh and Seleznev (2012). The model of active transport in some compartments of mammalian cells will be considered, and the possible mechanisms for regulating the concentrations of substances in these compartments will be discussed. Several models of these compartments will be considered in Chap. 3.

2.6.1 Mitochondria

The mitochondrion is one of the most complex cellular compartments that plays an important role in the life of a cell. First, the importance of a mitochondrion is determined by the fact that it transforms free energy from nutrients and oxygen into ATP, which is a universal energy carrier and is convenient to use in any part of a living organism. Because the work that is performed by animals in the environment is frequently accompanied by peak loads, ATP production should be controlled so that it is possible to change the ATP synthesis rate by many orders of magnitude during a short period. Therefore, the structure and functions of a mitochondrion possess a complex system of regulation, which is linked to the variation in Ca^{2+} concentration between the matrix of a mitochondrion and the cytoplasm. In particular, a change in Ca^{2+} concentration initiates the mechanisms of programmable death of a cell, and this change can lead to a change in the matrix volume.

A mitochondrion is covered by two membranes. The inner membrane consists of folds (cristae) that provide a maximum surface area. This membrane contains principal enzymes that provide a selective exchange of substances between the cellular plasma and the matrix. Furthermore, the matrix contains enzymes that participate in the Krebs cycle and fatty acid oxidation.

It is believed that during the early stages of evolution, mitochondria were likely to be an independent organism because its matrix contains ribosomes and DNA. Moreover, enzymes that are built into the inner membrane play a significant role in the processes that occur at this membrane (Beard and Qian 2008, Melkikh and Seleznev 2012). The scheme of the mitochondrion is shown in Fig. 2.22.

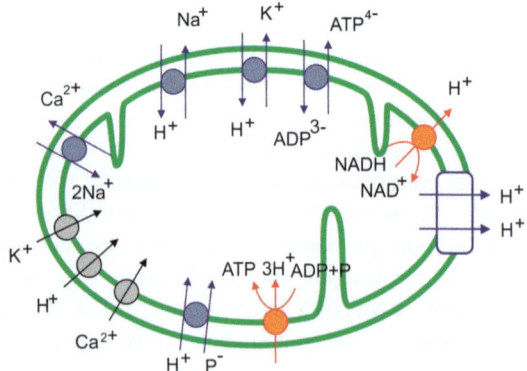

Fig. 2.22 Simplified scheme of the ion transporters in the inner mitochondrial membrane

2.6 Regulation of Ion Transport in Compartments of a Mammalian Cell

Details of the reactions that occur through the mitochondria enzymes are discussed in Melkikh and Seleznev (2012), and this section will present the main results from modeling ion transport through the membrane of the mitochondria.

We will consider the systems of the ion transport membrane of the mitochondria, which function independently of the respiratory chain and are not affiliated with the respiratory chain by stoichiometry integer coefficients because the respiratory chain was analyzed and shown previously (Melkikh and Seleznev 2012).

The ATPase enzyme maintains the ATP synthesis reaction at the expense of the free energy of protons $\Delta \mu_H^e$.

$$ADP + P + nH_i^+ + F \rightleftarrows ATP + nH_o^+ + F. \quad (2.48)$$

Therefore, the flux of the reaction is represented by the following equation:

$$J_A = k_{ATP} n_F n_A^i n_H^o \left(\exp(n\Delta\mu_H^e - \Delta\mu_A) - 1 \right). \quad (2.49)$$

where k_{ATP} is the reaction rate constant,

$$\Delta\mu_H^e = \ln \frac{n_H^o}{n_H^i} - e\varphi, \quad \Delta\mu_A = \ln \frac{n_A^i n_{D0}^i n_{P0}^i}{n_D^i n_P^i n_{A0}^i} \quad (2.50)$$

where $n_A^i, n_D^i, n_P^i, n_{A0}^i, n_{D0}^i, n_{P0}^i$ are the concentrations of ATP, ADP, and P inside a mitochondrion and their corresponding concentrations at equilibrium, respectively.

The next enzyme of the mitochondrion membrane is adenosine nucleotide translocase. This enzyme supplies the substrate for ADP^{3-} synthesis from the cytoplasm into the mitochondrion, and the reaction product, ATP^{4-}, travels through the membrane and into the cellular plasma. The equation for the ion exchange reaction through the antiporter under consideration is as follows:

$$ADP_o^{3-} + ATP_i^{4-} + F \rightleftarrows ADP_i^{3-} + F + ATP_o^{4-}. \quad (2.51)$$

The ion flow through the antiporter will take the following form:

$$J_{TD} = \frac{n_F}{2} \frac{k_{D\uparrow} k_{T\uparrow} (k_{T\downarrow} + k_{D\downarrow})}{k_{T\downarrow} k_{D\downarrow}} \left[n_D^o n_T^i \exp(\varphi) - n_D^i n_T^o \right], \quad (2.52)$$

where $k_{D\uparrow}, k_{T\uparrow}, k_{D\downarrow}, k_{T\downarrow}$ are the constants of ADP^{3-} and ATP^{4-} binding and release rate, respectively.

The K^+–H^+ and Na^+–H^+ antiporters are described by the following formulas:

$$J_{NaH} = \frac{n_F^{NaH}}{2} \frac{k_{Na\uparrow} k_{H\uparrow} (k_{Na\downarrow} + k_{H\downarrow})}{k_{H\downarrow} k_{Na\downarrow}} \left[n_{Na}^o n_H^i - n_{Na}^i n_H^o \right], \quad (2.53)$$

$$J_{KH} = \frac{n_F^{KH}}{2} \frac{k_{K\uparrow} k_{H\uparrow} (k_{K\downarrow} + k_{H\downarrow})}{k_{H\downarrow} k_{K\downarrow}} \left[n_K^o n_H^i - n_K^i n_H^o \right]. \quad (2.54)$$

A similar expression for the flow characterizes the Ca^{2+}–2Na$^+$ exchanger:

$$J_{CaNa} = \frac{n_F^{CaNa}}{2} \frac{k_{Na\uparrow} k_{Ca\uparrow}(k_{Na\downarrow} + k_{Ca\downarrow})}{k_{Ca\downarrow} k_{Na\downarrow}} \left[n_{Ca}^o (n_{Na}^i)^2 - n_{Ca}^i (n_{Na}^o)^2 \right]. \quad (2.55)$$

The flow through the P–H$^+$ symporter is described by the following formula:

$$J_{PH} = \frac{n_F^{PH} k_\uparrow^i k_\downarrow^o}{k_\downarrow^o + k_\uparrow^i} \left[n_P^o n_H^o - n_P^i n_H^i \right]. \quad (2.56)$$

The ion flows through the electrogenic H$^+$, K$^+$, Ca^{2+}, Cl$^-$ uniporters:

$$J_H = \frac{n_F^H k_{H\uparrow}^i k_{H\downarrow}^o}{k_{H\downarrow}^i} \left[n_H^o - \exp(\varphi) n_H^i \right], \quad (2.57)$$

$$J_K = \frac{n_F^K k_{K\uparrow}^i k_{K\downarrow}^o}{k_{K\downarrow}^i} \left[n_K^o - \exp(\varphi) n_K^i \right], \quad (2.58)$$

$$J_{Ca} = \frac{n_F^{Ca} k_{Ca\uparrow}^i k_{Ca\downarrow}^o}{k_{Ca\downarrow}^i} \left[n_{Ca}^o - \exp(2\varphi) n_{Ca}^i \right], \quad (2.59)$$

$$J_{Cl} = \frac{n_F^{Cl} k_{Cl\uparrow}^i k_{Cl\downarrow}^o}{k_{Cl\downarrow}^i} \left[n_{Cl}^o \exp(\varphi) - n_{Cl}^i \right]. \quad (2.60)$$

By using the flow of ions that was published previously (Melkikh and Seleznev 2012), the system of equations for the conservation of particles of each component and charge were recorded. This system allows the determination of the internal concentration and electrical potential of mitochondria as a function of the concentration in the solution and time.

Due to the complexity of the system of equations, the solution is possible only with the use of numerical methods. To accomplish the task of restoring the characteristics of the flow of ions through the mitochondrial membrane, must be utilized a special program of experimental work to the duplicate samples of mitochondrial suspensions from the cell body of a particular species to restore the features of the investigated organelles.

Melkikh and Seleznev (2012) undertook a simpler alternative approach, which allowed for the derivation of analytical formulas for the potential and internal concentrations of ions as functions of their external concentrations. It is suggested that through evolution, living organisms have selected structures and physiological processes that optimally control vital functions and provide the necessary adaptability and robustness of that control.

Based on this principal and assuming that the system has reached steady state, we can obtain the following system of equations:

2.6 Regulation of Ion Transport in Compartments of a Mammalian Cell

$$mJ_r + nJ_A =$$
$$= mA^* \left(\exp(\Delta\mu_r - m\Delta\mu_H^e) - 1\right) + n\left(\exp(n\Delta\mu_H^e - \Delta\mu_A) - 1\right) = 0 \tag{2.61}$$

$$n_T^i + n_D^i = n_T^o + n_D^o, \tag{2.62}$$

$$n_T^i n_D^o \exp(e\varphi) - n_D^i n_T^o = 0, \tag{2.63}$$

$$n_P^o n_H^o - n_P^i n_H^i = 0, \tag{2.64}$$

$$n_K^i n_H^o - n_K^o n_H^i = 0, \tag{2.65}$$

$$n_{Cl}^o \exp(\varphi) - n_{Cl}^i = 0, \tag{2.66}$$

$$n_n^i + n_{Cl}^i + 2n_T^i + n_D^i + n_P^i = n_K^i + n_H^i. \tag{2.67}$$

Here, A* is a constant that relates to the ratio of the numbers of F-ATPases and respiratory chains. In the obtained system, the following values are unknown: μ_H^e, n_T^i, n_D^i, n_P^i, n_K^i, n_{Cl}^i, φ, and all of these variables should be determined by the seven Eq. (2.61–2.67).

In accordance with Melkikh and Seleznev (2012), we shall now consider the normal physiological condition. For ATP synthesis ($J_A > 0$), the difference between chemical potentials of the synthesis reaction $\Delta\mu_A^i$ was less than $n\Delta\mu_H^e$. When utilizing the experimental value of $\Delta\mu_A^i = 25.6$, it is easy to determine the stoichiometric number n:

$$n \geq \frac{25.6}{8.2} \approx 3.$$

Taking into consideration that

$$\mu_H^e = \ln\frac{n_H^o}{n_H^i} - \varphi, \tag{2.68}$$

the following expression for the ratio of proton concentrations inside and outside a mitochondrion can be easily derived:

$$\frac{n_H^o}{n_H^i} = \exp\left(\frac{\mu_A^i}{3} + \varphi\right). \tag{2.69}$$

$$n_T^i = \frac{(n_T^o + n_D^o)n_T^o \exp(-\varphi)}{n_T^o + n_D^o \exp(-\varphi)}. \tag{2.70}$$

$$n_D^i = n_T^o + n_D^o - \frac{(n_T^o + n_D^o)n_T^o \exp(-\varphi)}{n_T^o + n_D^o \exp(-\varphi)} = \frac{n_D^o(n_T^o + n_D^o)}{n_T^o + n_D^o \exp(-\varphi)}. \tag{2.71}$$

Phosphate ion concentration:

$$n_P^i = \frac{n_P^o n_H^o}{n_H^i} = n_P^o \exp\left(\frac{\Delta\mu_A^i}{3} + \varphi\right). \tag{2.72}$$

Potassium ion concentration:

$$n_K^i = \frac{n_K^o n_H^i}{n_H^o} = n_K^o \exp\left(-\frac{\Delta\mu_A^i}{3} - \varphi\right). \tag{2.73}$$

By using the neutrality equation, we shall obtain the approximate equation for determining the resting potential:

$$\exp(-\varphi) = \frac{2(n_T^o + n_D^o) + n_n^i}{(n_K^o + n_H^o)} \exp\left(\frac{\Delta\mu_A^i}{3}\right), \text{ or} \tag{2.74}$$

$$\varphi = \ln\frac{(n_K^o + n_H^o)}{2(n_T^o + n_D^o) + n_n^i} - \frac{\Delta\mu_A^i}{3}. \tag{2.75}$$

A simplified system of equations allows us to understand the basic mechanisms that are responsible for the reaction in the mitochondria after an environmental change.

Another area of mitochondrion physiology investigation is related to the study of regulatory functions of calcium transport systems. As shown by studies from the last 30 years, the cell utilizes a large sum of energy in maintaining a low concentration of calcium ions. Calcium ions are universal intracellular regulators, they transmit incoming signals to the cell through enzymatic systems. In the resting cell, the concentration of calcium is low, but when the cell receives the appropriate signal, it responds with an avalanche-like increase in the concentration of calcium ions.

It is necessary that cellular systems control the concentration of calcium in the cytoplasm and are able to rapidly lower these levels, and these systems are built into the cell membrane. In the outer membrane, the Ca-ATPase is a pump that acts against the Ca^{2+} gradient from the cell and into the intercellular medium. An additional system, the Ca^{2+}–Na^+ exchanger, is responsible for lowering the concentration of calcium in the cytoplasm by exchanging intracellular Ca^{2+} for extracellular Na^+.

The Ca-ATPase is located on the membrane of the endoplasmic reticulum. This enzyme pumps Ca^{2+} ions from the cytoplasm into the endoplasmic reticulum cisterns by way of ATP hydrolysis. In addition, in the mitochondria, a special transport system can pump calcium from the cytoplasm to the matrix.

Increased intracellular Ca^{2+} is a type of alarm. In response to an increase in the concentration of calcium ions, the cell mobilizes all its systems, which removes calcium. Furthermore, the elevated concentration of calcium in the cell occurs only for a short period, which allows for transmission of the stimulus.

The above scheme of signal transduction through calcium ions is at the base of many systems, especially of muscle contraction.

2.6.2 Sarcoplasmic and Endoplasmic Reticulum

The sarcoplasmic reticulum (SR) is a depot for calcium ions. To provide this function, the SP membrane contains an ATP-dependent calcium pump that transports two calcium ions from the cellular plasma into the SP for one event of ATP hydrolysis.

The SR membrane is connected to the tubes of the T-system, which are directly linked to the membrane of a muscular cell, by special membrane bridges (Fig. 2.23). The signal for necessary muscle contraction travels through this system to the SR membrane, Ca^{+2} channels open, and Ca^{+2} ions are injected into the cytoplasm of the muscle cell where they excite muscle fiber contraction. During the rest period, Ca^{+2} channels close, and Ca^{+2} ions are pumped into cisterns by an ATP-dependent pump. For the described system to function, potential-dependent, passive channels with variable permeability and an active calcium pump should be present in the SP membrane.

The results of previous investigations (Meissner 2001; Shannon et al. 2000, 2001) show that no electric potential exists on the SP membrane or that this potential is very low. Moreover, when the channels are closed, Ca^{+2} pumps maintain the ratio of the concentration of Ca^{+2} ions inside and outside the SP at approximately 7,000. It is reported that the Ca^{+2} pump also works as the mode of ATP synthesis.

Functions of the endoplasmic reticulum (ER) consist of accumulation and transport of substances that are important for a cell, and the ER, like the SR, is a depot for Ca^{+2}. The resting potential on the ER membrane is obtained within the framework of the model for ion transport in the ER that was developed in a previous paper (Marhl et al. 1998), and the balance of Ca^{+2} concentration in the cytosol is carried out within this model. In particular, the passive and active flows of Ca^{+2} ions through the ER membrane contribute substantially to this balance.

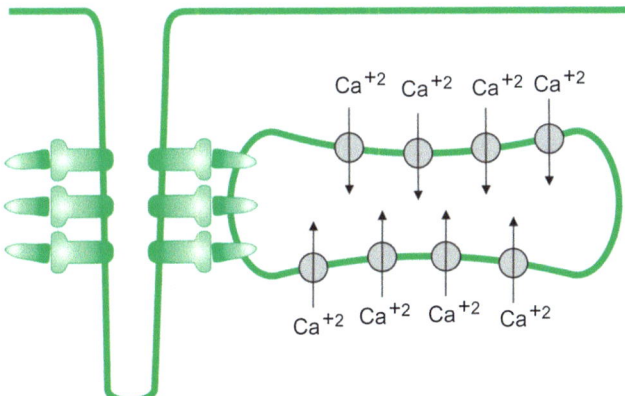

Fig. 2.23 Simplified scheme of the ion transporters in sarcoplasmic reticulum

Thermodynamically correct expressions for the ratio of the concentrations of calcium ions on the membranes of the SR and ER were obtained previously (Melkikh and Seleznev 2012) on the basis of the proposed algorithm. From experimental data on the stoichiometry of the ATP-dependent calcium pump (two Ca^{2+} ions for the hydrolysis of one molecule of ATP), the following reaction can be written:

$$F + 2Ca_o^{2+} + T \Leftrightarrow F + 2Ca_i^{2+} + D + P. \quad (2.76)$$

In accordance with (6.6), we write the expression for the calcium flux with a small calcium concentration as:

$$J_{Ca}^{ATP} = k_\downarrow^{Ca} n_F n_D n_P \left(n_{Ca}^i\right)^2 \exp(4\varphi)[\exp(\Delta\mu_A - 4e\varphi - 2\Delta\mu_{Ca}) - 1] \quad (2.77)$$

Because the calcium concentration is much smaller than the concentrations of potassium, sodium and other ions, and because there are no pumps other than calcium pump, in the SR, the potential will be close to zero. This result is in agreement with the experimental data. Then, following from (2.77), the ratio of the concentrations of calcium ions across the membrane in the case ($J_{Ca}^{ATP} = 0$) can be determined by the relation:

$$\Delta\mu_{Ca} = \ln\frac{n_{Ca}^i}{n_{Ca}^o} = \frac{\Delta\mu_A}{2}. \quad (2.78)$$

With $\Delta\mu_A = 20$, we have:

$$\frac{n_{Ca}^i}{n_{Ca}^o} = \exp(10) \approx 22000,$$

which is in qualitative agreement with the experimental value of this ratio. The reason for the difference, in all likelihood, is that we have neglected passive transport of Ca^{2+} ions in (2.78).

2.6.3 Synaptic Vesicles

A model of active ion transport in synaptic vesicles has been discussed in detail in Melkikh and Seleznev (2007, 2012). Here, we consider the model's basic results.

Neurotransmitters are molecules that open channels and control the generation of nerve impulses in neurons (for example, see Nicholls et al. 2001). At rest, neurotransmitters are kept in synaptic vesicles (or in large, dense-core vesicles and synaptic-like microvesicles (Harada et al. 2010)). During impulse propagation, neurotransmitters leave the vesicles and open neighboring cellular channels in the place of synaptic contacts.

According to some publications (Nicholls et al. 2001), a proton gradient, which is created by H-ATPase as it transports protons into synaptic vesicles, is used to transport neurotransmitters into synaptic vesicles.

2.6 Regulation of Ion Transport in Compartments of a Mammalian Cell

Experiments have shown that proton flow into a synaptic vesicle is provided by energy from ATP. An expression for the flux of protons through an H-ATPase has the following form:

$$J_H = \tilde{C}_H \left[\exp(\Delta \mu_A)(n_H^o) - \exp(4\varphi)(n_H^i) \right] \tag{2.79}$$

From this equation, we can calculate the steady-state proton concentration within a synaptic vesicle as:

$$\exp\left(\frac{\Delta \mu_A}{4} - \varphi\right) n_H^o = n_H^i. \tag{2.80}$$

Neurotransmitters accumulate in various synaptic vesicles with individual active transport systems. The schemes of select neurotransmitter transport systems are shown in Figs. 2.24, 2.25, 2.26 (Nicholls et al. 2001).

The passive flow of all neurotransmitters was neglected based on the need for the maximum effectiveness of ATP hydrolysis in energy utilization.

The active flow of neutral neurotransmitters into a vesicle has the following form for monoamines and acetylcholine:

$$J_T = C_1 \left(n_T^o \exp(\varphi) (n_H^i)^2 - n_T^i (n_H^o)^2 \right). \tag{2.81}$$

Expressing this equation as the internal concentration of positive neurotransmitters with an electroneutrality condition, the resting potential can be expressed as:

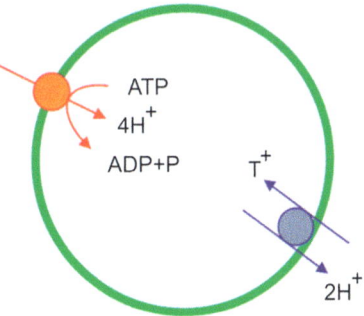

Fig. 2.24 The transport scheme of monoamines and acetylcholine

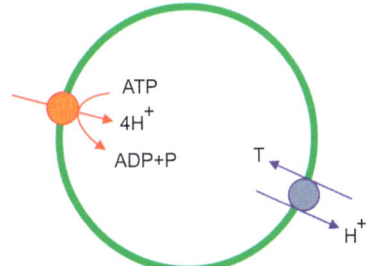

Fig. 2.25 The transport scheme of GABA and glycine

Fig. 2.26 The transport scheme of glutamate

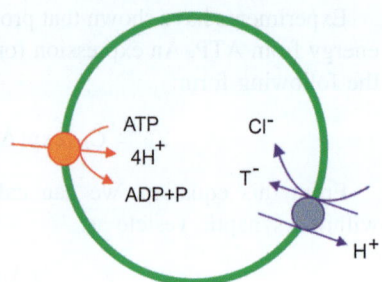

$$\varphi = \ln\left(\frac{n_T^i + \sqrt{(n_T^i)^2 + 4n_-^o n_+^o}}{2n_-^o}\right). \quad (2.82)$$

The concentrations of passive negative ion and positive ions are roughly equal on the outside of the vesicle and are on the order of 0.1 M. The concentration of the mediator on the inside of the synaptic vesicle is approximately 0.5 M. Considering that $\Delta\mu_A$ has a magnitude of 20, the calculated potential is approximately 41 mV.

The proton concentration within the vesicle is equal to 0.0011 mM, which corresponds to a ΔpH of 1.46 across the vesicle membrane. This value coincides with the experimental data, which reports a $\Delta pH < 2$ (Gidon and Sihra 1989; Tabb et al. 1992).

The mediator concentration outside of the vesicle is approximately 0.1 mM.

Similarly, we can derive an expression for the potential of GABA and glycine as follows:

$$\varphi = \frac{1}{2}\ln\frac{n_H^o \exp\left(\frac{\Delta\mu_A}{4}\right) + n_K^o}{n_{Cl}^o} \approx \frac{n_H^o \exp\left(\frac{\Delta\mu_A}{4}\right)}{2n_{Cl}^o} \approx 2.97 \times 10^{-5}. \quad (2.83)$$

In dimensional units, the potential is 7.4×10^{-4} mV. This value is very small in comparison to the characteristic potentials of a cellular membrane. In this case we can consider the potential on the membrane of a synaptic vesicle to be absent.

The cellular mediator concentration is 3.4 mM in this case. This value qualitatively agrees with the experimental data.

For glutamate transport in a stationary case, the expression for potential will take the form:

$$\varphi = \ln\left(\frac{-n_T^i + \sqrt{(n_T^i)^2 + 4n_-^o n_+^o}}{2n_-^o}\right), \quad (2.84)$$

which results in a negative potential of approximately—41 mV in dimensional units. The cellular neurotransmitter concentration is 18 mM. This result is also on the same order of magnitude as the experimental data.

Neurotransmitters play an extremely significant role in ensuring the normal functioning of the nervous system. At the same time, maintaining the concentration of neurotransmitters in the intercellular space and regulating the transport of synaptic vesicles within the cell are also important. However, models of these processes are absent from the literature.

2.7 Conclusions

By constructing models of transport systems in four cell types and compartments, we were able to verify some of the proposed algorithms in Chap. 1. A comparison with experimental data suggested that the proposed algorithms can qualitatively describe ion transport processes across cellular membranes and compartments. A transport system model was constructed for steady-state, allowing us to offer algorithms for the regulation of ion transport. Ion concentrations could be maintained at level required by the cells, which may correspond to conditions with insufficient experimental data.

Good agreement between the numerical modeling results and the mammalian cell data supports the validity of our previous hypotheses to simplify the model. In particular, this result supports the electroneutrality of the internal environment and the minimum osmotic pressure difference across the membrane. Thus, within the framework of our algorithm, it can be argued that animal cells have sufficient commonality and that their structural and functional differences can be ignored. In the future, the proposed algorithm can be used for building regulatory transport systems in any animal cell, even in the absence of experimental data. Also justified is the assumption that, in the framework of our model, compartments can be considered isolated systems for modeling transport across the membrane.

References

Alvarez-Leefmans FJ (2001) Intracellular chloride regulation. In: Sperelakis N (ed) Cell physiology sourcebook, 3rd edn. Academic Press, San Diego

Barish ME (1991) Increases in intracellular calcium ion concentration during depolarization of cultured embryonic xenopus spinal neurons. J Physiol 444:545–565

Baumgarten CM, Feher JI (2001) Osmosis and regulation of cell volume. In: Sperelakis N (ed) Cell physiology sourcebook, 3rd edn. Academic Press, San Diego

Beard DA, Qian H (2008) Chemical biophysics quantitative analysis of cellular systems. Cambridge University Press, Cambridge

Cavalier-Smith T (1998) A revised six-kingdom system of life. Biol Rev Camb Philos Soc 73(3):203–266

Cavalier-Smith T (2002) The phagotrophic origin of eukaryotes and phylogenetic classification of protozoa. Int J Syst Evol Microbiol 52:297–354

Chesler M (2003) Regulation and modulation of pH in the brain. Physiol Rev 83:1183–1221

Craciun G, Brown A, Friedman A (2005) A dynamical system model of neurofilament transport in axons. J Theor Biol 237:316–322

El-Mallakh RS (2004) Ion homeostasis and the mechanism of action of lithium. Clin Neurosci Res 4:227–231

Faber GM, Rudy Y (2000) Action potential and contractility changes in $[Na^+]_i$ overloaded cardiac myocytes: a simulation study. Biophys J 78:2392–2404

Fossat B, Porthé-Nibelle J, Pedersen S, Lahlou B (1997) Na^+/H^+ exchange and osmotic shrinkage in isolated trout hepatocytes. J Exp Biol 200:2369–2376

Freedman JC (2001) Membrane transport in red blood cell. In: Sperelakis N (ed) Cell physiology sourcebook, 3rd edn. Academic Press, San Diego

Freedman JC, Hoffman JF (1979) Ionic and osmotic equilibria of human red blood cells treated with nystatin. J Gen Physiol 74:157–185

Freudenrich CC, Murphy E, Liul S, Lieberman M (1992) Magnesium homeostasis in cardiac cells. Mol Cell Biochem 114(1–2):97–103

Furimsky M, Moon TW, Perry SF (2000) Evidence for the role of a Na^+/HCO_3^- cotransporter in trout hepatocyte pHi regulation. J Exp Biol 203:2201–2208

Garcia-Romeu F, Cossins AR, Motais R (1991) Cell volume regulation by trout erythrocytes: characteristics of the transport systems activated by hypotonic swelling. J Physiology 440:547–567

Gidon S, Sihra T (1989) Characterization of a H^+-ATPase in rat brain synaptic vesicles. Coupling to l-glutamate transport. J Biol Chem 264(14):8281–8288

Giffard RG, Papadopoulos MC, van Hooft JA, Xu L, Guiffrida R, Monyer H (2000) The electrogenic sodium bicarbonate cotransporter: developmental expression in rat brain and possible role in acid vulnerability. J Neurosci 20(3):1001–1008

Gusev GP, Agalakova NI, Lapin AV (1996) Activation of the Na^+-K^+ pump in frog erythrocytes by catecholamines and phosphodiesterase blockers. Biochem Pharmaco 52(9):1347–1353

Harada K, Matsuoka H, Nakamura J, Fukuda M, Inoue M (2010) Storage of GABA in chromaffin granules and not in synaptic-like microvesicles in rat adrenal medullary cells. J Neurochem 114:617–626

Hoffman EK, Dunham PB (1995) Membrane mechanisms and intracellular signaling in cell volume regulation. Int Rev Cytol 161:173–262

Hume JR, Duan D, Coller ML, Yamazaki J, Horowitz B (2000) Anion transport in heart. Physiol Rev 80(1):31–81

Jamshidi N, Edwards JS, Fahland T, Church GM, Palsson BO (2001) Dynamic simulation of the human red blood cell metabolic network. Bioinform 17(3):286–287

Jamshidi N, Wiback S, Palsson BO (2002) In silico model-driven assessment of the effects of single nucleotide polymorphisms (SNPs) on human red blood cell metabolism. Genome Res 12(11):1687–1692

Joshi A, Palsson BO (1989) Metabolic dynamics in the human red cell: part I. a comprehensive kinetic model. J Theor Biol 141(4):515–528

Lew VL, Bookchin RM (1986) Volume, pH and ion-content regulation in human red cells: analysis of transient behavior with an integrated model. J Membr Biol 92:57–74

Lluch MM, de la Sierra A, Poch E, Coca A, Aguilera MT, Compte M, Urbano-Marquez A (1996) Erythrocyte sodium transport, intraplatelet pH, and calcium concentration in salt-sensitive hypertension. Hypertens 27(4):919–925

Luo Ch-H, Rudy Y (1991) A model of the ventricular cardiac action potential, depolarization, repolarization and their interaction. Circ Res 68:1501–1526

Luo Ch-H, Rudy Y (1994) A dynamic model of the cardiac ventricular action potential. Circ Res 74(6):1071–1096

MacManus ML, Churchwell KB, Strange K (1995) Regulation of cell volume in health and disease. N Engl J Med 9:1260–1266

Marhl M, Schuster S, Brumen M, Heinrich R (1998) Modelling oscillations of calcium and endoplasmic reticulum transmembrane potential. Role ot the signaling and buffering proteins and of the size Ca^{2+} sequestering ER subcompartments Bioelectroch Bioener 46(1):79–90

Meissner G (2001) Ca_2^+ release from sarcoplasmic reticulum. In: Sperelakis N (ed) Cell physiology sourcebook, 3rd edn. Academic Press, San Diego

Melkikh AV, Seleznev VD (2007) Models of active transport of neurotransmitters in synaptic vesicles. J Theor Biol 248(2):350–353

Melkikh AV, Seleznev VD (2012) Mechanisms and models of the active transport of ions and the transformation of energy in intracellular compartments. Prog Biophys Mol Bio 109(1–2):33–57

Melkikh AV, Sutormina MI (2008) Model of active transport of ions in cardiac cell. J Theor Biol 252(2):247–254

Melkikh AV, Sutormina MI (2010) A model of active transport of ions in hepatocytes. Biophys 55(1):67–70

Mulquiney PJ, Bubb WA, Kuchel PW (1999) Model of 2,3-bisphosphoglycerate metabolism in the human erythrocyte based on detailed enzyme kinetic equations: in vivo kinetic characterization of 2,3-bisphosphoglycerate synthase/phosphatase using 13C and 31P NMR. Biochem J 342(3):567–580

Murphy E, Coll K, Rich TL, Williamson JR (1980) Hormonal effects on calcium homeostasis in isolated hepatocytes. J Biol Chem 255:6600–6608

Murphy E (2000) Mysteries of magnesium homeostasis. Circ Res 86(3):245–248. doi:10.1161/01.RES.86.3.245

Nicholls JG, Martin AR, Wallace BG, Fuchs PA (2001) From neuron to brain, 4th edn. Sinauer Associates, Sunderland

Raupach T, Ballanyi K (2004) Intracellular pH and K_{ATP} channel activity in dorsal vagal neurons of juvenile rats in situ during metabolic disturbances. Brain Res 1017:137–145

Sarkadi B, Parker JC (1991) Activation of ion transport pathways by changes in cell volume. Biochim Biophys Acta 1071(4):407–427

Shannon TR, Chu G, Kranias EG, Bers DM (2001) Phospholamban decrease the energetic efficiency of the sarcoplasmic reticulum ca pump. J Biol Chem 276(10):7195–7201

Shannon TR, Ginzburg KS, Bers DM (2000) Reverse mode of the sarcoplasmic reticulum calcium pump and load-dependent cytosolic calcium decline in voltage-clamped cardiac ventricular myocytes. Biophysical J 78:322–333

Sillau AH, Escobales N, Juarbe C (1996) Differences in membrane ion transport between hepatocytes from the periportal and the pericentral areas of the liver lobule. Experientia 52:554–557

Sperelakis N (2000) Physiology and Pathophysiology of the Heart, 4th edn. Academic Publishers, Boston

Sperelakis N, Gonzales-Serratos H (2001) Skeletal muscle action potentials. In: Sperelakis N (ed) Cell physiology sourcebook, 3rd edn. Academic Press, San Diego

Sriboonlue P, Jaipakdee S, Jirakulsomchok D, Mairiang E, Tosukhowong P, Prasongwatana V, Savok S (2005) Changes in erythrocyte contents of potassium, sodium and magnesium and Na, K-pump activity after the administration of potassium and magnesium salts. J Med Assoc Thai 87(12):1506–1512

Swietach P, Tiffert T, Mauritz JMA, Seear R, Esposito A, Kaminski CF, Lew VL, Vaughan-Jones RD (2010) Hydrogen ion dynamics in human red blood cells. J Physiol 588:4995–5014

Tabb JS, Kish PE, Van Dyke R, Ueda T (1992) Glutamate transport into synaptic vesicles. roles of membrane potential, pH gradient and intravesicular pH. J Biol Chem 267(22):15412–15418

Werner A, Heinrich R (1985) A kinetic model for the interaction of energy metabolism and osmotic states of human erythrocytes. Analysis of the stationary "in vivo" state and of time dependent variations under blood preservation conditions. Biomed Biochim Acta 44(2):185–212

Yachie-Kinoshita A, Nishino T, Shimo H, Suematsu M, Tomita M (2010) A metabolic model of human erythrocytes: practical application of the E-cell simulation environment. J Biomed Biotechnol ID 642420. doi:10.1155/2010/642420

Chapter 3
Models of Ion Transport and Regulation in Plant Cells and Unicellular Organisms

Models of ion transport in microorganisms, plants and their compartments have been considered. A model of the regulation of ion transport in changing extracellular concentrations has also been constructed. In contrast to animal cells, microorganisms can survive when a large concentration gradient exists between the inside and the outside of the cell and when significant ion concentration changes occur in the environment.

3.1 Introduction

In this chapter, we consider a model of active ion transport in the cells of simpler organisms, namely, prokaryotic cells, plant cells, fungi and their compartments. Prokaryotic cells, in contrast to eukaryotic cells, do not have a formalized nucleus or other internal membranous organelles. Prokaryotes consist of bacteria, including cyanobacteria and archaea. It is believed that the descendants of prokaryotic cells are the organelles of eukaryotic cells, such as mitochondria and chloroplasts. Some prokaryotic and eukaryotic cells fall under the category of "microorganisms".

The ubiquity and the total metabolic potential of microorganisms determine their crucial role in substance circulation and in the maintenance of a dynamic equilibrium in Earth's biosphere. For example, diatoms constitute approximately one quarter of the Earth's total biomass. Small, free-swimming algae are a part of plankton, causing "blooms" of water when they develop in large quantities.

Some microorganisms are able to grow in extreme conditions. For example, some yeast and archaea can grow in a NaCl solution that is close to saturation. Some archaea are acidophilic and can live in acidic environments (pH 1–5), whereas alkaliphilic archaea prefer an alkaline medium (pH 9–11).

A mammalian cell's turgor, which is the difference between its internal and external pressures, is less than one atmosphere, whereas plants and fungi have much higher cellular turgor. Typically, the internal pressure in plant and fungal cells ranges from 5 to 10 atmospheres. The difference in the internal and external cellular pressures can reach 50 or even 100 atmospheres in fungi or in plants that grow in saline soils.

All of the above characteristics lead to significant differences between ion transport in microorganisms and plant versus animal cells.

A distinctive feature of microorganisms is their ability to survive in various environments, which essentially differ depending upon their composition (sometimes by orders of magnitudes in particular ion concentrations). Microorganisms use various strategies for this purpose. One of their major strategies is the regulation of ion transport, which consists of switching on (switching off) or increasing (decreasing) the intensity of their transport systems in response ion concentration changes in the environment. The most important transport systems in microorganisms, which are studied extensively, are the pH system and the regulation of NaCl (salt stress). At the same time, despite an ample quantity of experimental data, theoretical models of the regulation of ion transport are absent from the literature, and the only models of ion transport in microorganisms (that do not account for regulation) are the works of the author (Melkikh and Seleznev 2009; Melkikh and Bessarab 2010). An absence of regulatory models limits our ability to reveal any general laws of regulation strategies and prevents us from predicting how microorganisms (including artificial microorganisms) will behave in extreme conditions (for example, with an essential lack or surplus of any ion).

In the articles (Melkikh and Seleznev 2009; Melkikh and Bessarab 2010), models of ion transport in archaeabacteria and in diatoms were constructed. Let us revisit these models here.

3.2 Archaea

A model of active transport in archaeabacteria was discussed in detail in the article (Melkikh and Seleznev 2009).The article also discussed the notable features of archaea. These features include the following:

1. A notable feature is their unusual ribosomal RNA and transport RNA. Distinctive features were also identified in other components of their protein synthesis system.
2. Unlike all other organisms, archaeal membrane lipids include polyatomic alcohols rather than fatty acids, making their membranes especially strong.
3. The shells of some archaeal cells have surface layers that are made of various structured, regularly arranged protein or glycoprotein molecules, further strengthening their membranes.
4. The majority of archaea are extremophiles, i.e. they develop under extreme conditions, such as elevated temperatures, high acidity and saturated salt solutions.

3.2 Archaea

It was also noted that halophilic archaea live in highly concentrated solutions, such as a NaCl concentration of 20 % or greater.

The cell and its transporters are schematized in Fig. 3.1.

Active transport systems in the *Halobacterium* genus of extremely halophilic archaea include the following:

- H^+-pump (ATP-dependent proton pump);
- Na^+–H^+-exchanger independent of ATP (Na^+/H^+-antiporter);
- H^+–K^+-pump (ATP-dependent);
- Na^+–Ca^{2+}-exchanger independent of ATP;
- Mg^{2+}–H^+-exchanger independent of ATP.

Moreover, ATP is synthesized on the cell membrane due to the electrochemical gradient of protons, which is the result of the electronic respiratory chain and the absorption of light by bacteriorhodopsin.

The proton flux is written in accordance with Chap. 1 as follows:

$$J_H = C_H \left[\exp(\Delta \mu_A + n\varphi)(n_H^i)^n - (n_H^o)^n \right] = 0 \quad (3.1)$$

For *Halobacterium salinarum* cells, it was found that the stoichiometry of the H^+-ATPase pump results in the transfer of three protons to the environment during the hydrolysis of one ATP molecule (Oren 1999). The formula for the H^+-ATPase (3.1) then becomes:

$$J_H = C_H \left[\exp(\Delta \mu_A + 3\varphi)(n_H^i)^3 - (n_H^o)^3 \right] = 0. \quad (3.2)$$

Considering the relevant experimental data, we assume $\Delta pH = 0.7$ for the *Halobacterium* genus of archaeal cells (Lanyi 1978).

If three protons are transferred to the environment, we have the following result:

$$\frac{n_H^i}{n_H^o} = \exp\left(-\frac{\Delta \mu_A}{3} - \varphi\right) = 0.188. \quad (3.3)$$

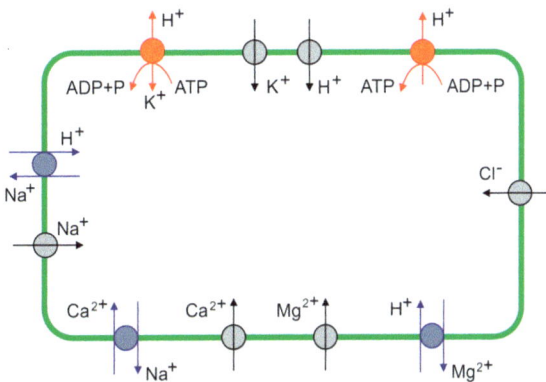

Fig. 3.1 The scheme of ionic membrane transport in the *Halobacterium* genus of archaeal cells. *Shaded* and *clear geometries* denote active and passive fluxes, respectively

The experimental ratio of the proton concentrations in the cell and in the environment is:

$$\frac{n_H^i}{n_H^o} = 0.190. \tag{3.4}$$

The calculated ratio of proton concentrations agrees well with the corresponding experimental ratio. This is an indication that the H⁺-ATPase pump is the major mechanism responsible for proton transport, while the passive transport of protons and the participation of these ions in other transport mechanisms are secondary to the H⁺-ATPase.

According to experimental data (Oren 1999; Schafer et al. 1999), sodium ions are transported in the *Halobacterium* genus of archaeal cells by means of an ATP-independent Na⁺–H⁺-exchanger (Na⁺/H⁺-antiporter).

The main physiological functions of the Na⁺–H⁺-exchanger are the regulation of intracellular pH, the maintenance of osmotic pressure within the cell, and the protection of the cell from cytoplasmic acidification. This exchanger pumps sodium ions out of the cell. The sodium difference of the electrochemical potential $\Delta\mu_{Na}^+$ is formed due to the proton difference of the potential $\Delta\mu_H^+$. Three stoichiometries are proposed in the literature (Oren 1999) for archaeabacterial cells: 1H⁺/1Na⁺, 2H⁺/1Na⁺, and 3H⁺/2Na⁺. The experimental and calculated ratios of the sodium ion concentrations on the inside and the outside the cell are in qualitative agreement if the first exchange ratio is used in the calculation. Thus, the exchanger stoichiometry is 1H⁺/1Na⁺.

Let us write the equation for the ion flux through the Na⁺–H⁺-exchanger, considering that one sodium ion is exchanged when one proton is withdrawn from the cell, as follows:

$$J_{Na-H} = C_{Na-H}\left(n_{Na}^i n_H^o - n_{Na}^o n_H^i\right) = 0, \tag{3.5}$$

Neglecting the passive transport of sodium ions through the membrane, equating the sodium ion flux (3.5) to zero and using formula (3.3) for calculation of the proton concentration at $n = 3$, we have:

$$\frac{n_{Na}^i}{n_{Na}^o} = \exp\left(-\frac{\Delta\mu_A}{3} - \varphi\right) = 0.20. \tag{3.6}$$

The experimental ratio of the sodium ion concentrations inside of the cell and in the environment is:

$$\frac{n_{Na}^i}{n_{Na}^o} = 0.32.$$

Thus, the similarity of the experimental and theoretical concentration ratios suggests that the Na⁺–H⁺-exchanger is the major mechanism responsible for sodium ion transport.

3.2 Archaea

Potassium ions are the main intracellular ions of halobacteria. Their extracellular to intracellular concentration ratio can be as large as 1:1000. Along with other ions, K^+ ions are necessary for maintaining an ionic equilibrium on the outside and the inside of the cell and for the stabilization of enzymes, membranes and other cellular structures.

The cellular membrane of extremely halophilic archaea is highly penetrable by potassium ions. There is evidence in the literature that *Halobacterium* cells are characterized by a multiplicative K^+-transport system. The transport system includes a K^+ accumulation system and uses the transmembrane potential difference and an ATP-dependent potassium pump for ion transport. According to Oren (1999), the pump is a H^+–K^+-pump that depends on ATP (to regulate pump activity) and exchanges cellular protons for potassium ions from the environment. The operation of this pump is analogous to Trk of the K^+-transport system in *E. coli* cells. The stoichiometry of this pump is known; it transports ions in a H^+:K^+ ratio of 2:1.

By analyzing the above information, it is possible to construct an equation describing the operation of the proton-potassium pump. We shall use the data for the resulting ionic flux produced by the transport ATPase to write the following:

$$J_{H-K} = C_{H-K} \left[\exp(\Delta\mu_A + \varphi)(n_H^i)^2 n_K^o - (n_H^o)^2 n_K^i \right] = 0, \qquad (3.7)$$

where C_{H-K} is a constant related to the transport of potassium ions into the cell and the transport of protons out of the cell.

However, the H^+–K^+-pump is not the major mechanism in archaeabacterial cells. This was confirmed by calculation of the ion concentrations on both sides of the membrane. Expressing the ratio from (3.7), we have:

$$\frac{n_K^i}{n_K^o} = \exp\left(-\frac{\Delta\mu_A}{3}\right) = 0.0013.$$

The calculated concentration ratio contradicts the corresponding experimental data, which indicates that the intracellular concentration of potassium is $\sim 10^2$ larger than its extracellular concentration in extremely halophilic archaeabacteria.

Note that in accordance with the regulatory ideology discussed in Chap. 1, the pump may be involved in the regulation of potassium ions when there is a deficiency of potassium in the medium.

Most probably, the main potassium transport system in archaeal cells is passive transport. Indeed, the membrane of *Halobacterium salinarum* cells is highly permeable to K^+ ions (Oren 1999). In this case,

$$n_K^i = n_K^o \exp(-\varphi). \qquad (3.8)$$

If $\varphi = 130$ mV (5.2 in dimensionless units), the n_K^i/n_K^o ratio is 148. The experimental ratio of the intracellular and extracellular potassium concentrations in *Halobacterium salinarum* cells is 160 units, confirming its passive distribution. Thus, this is the major mechanism of potassium transport.

Potassium supports the structural and functional integrity of membranes and participates in cellular signaling processes.

The Na^+–Ca^{2+}-exchanger is responsible for the active flux of potassium ions (Bogomolni 1977). The direction of transmembrane transfer of Na^+ and Ca^{2+} is determined by the ratios of their concentration differences on both sides of the membrane.

If more than two sodium ions are exchanged for one potassium ion, the drop in the transmembrane potential (membrane depolarization) will lead to an increase in the admission of Ca^{2+} and the withdrawal of Na^+ or a decrease in the withdrawal of Ca^{2+} and the admission of Na^+. From a functional viewpoint, the Na^+–Ca^{2+}-exchanger can operate differently; it can accumulate Ca^{2+} ions inside the cell and eject Ca^{2+} ions from the cell.

The exchanger is characterized by the following exchange ratios: $2Na^+/1Ca^{2+}$, $3Na^+/2Ca^{2+}$, and $3Na^+/1Ca^{2+}$. The Na^+–Ca^{2+}-exchanger with a stoichiometry of $3Na^+/1Ca^{2+}$ occurs most frequently in the cells of extremely halophilic archaea. The corresponding flux equation is written as:

$$J_{Na-Ca} = C_{Na-Ca}\left[n_{Ca}^i(n_{Na}^o)^3 - \exp(\varphi)n_{Ca}^o(n_{Na}^i)^3\right] = 0, \quad (3.9)$$

Let us determine the ratio of the calcium ion concentrations. Expressing n_{Ca}^i/n_{Ca}^o from (3.9) and using expression (3.6) for the ratio of the sodium concentrations, we have:

$$\frac{n_{Ca}^i}{n_{Ca}^o} = \exp(-\Delta\mu_A - 2\varphi) = 4.5 \times 10^{-5}$$

This calculated value is in satisfactory agreement with the relevant experimental data (the concentration ratio is 22×10^{-5}).

Magnesium is a universal regulator of the physiological and biochemical processes that take place within a cell.

Magnesium ions primarily regulate the operation of ATP-dependent exchangers. Intra- and extra-cellular magnesium participates in regulation of the concentrations and movement of calcium, potassium, sodium and phosphate ions. Magnesium interacts with cellular lipids and ensures the intactness of the cellular membrane.

A study of the accumulation and consumption of reserve magnesium phosphate in extremely halophilic archaeal cells (Smirnov et al. 2002) suggested that Mg^{2+} transport into *Halobacterium salinarum* cells is energy dependent.

We assume that the Mg^{2+}–H^+-exchanger that carries magnesium ions into *Halobacterium salinarum* cells has an operating mechanism that is analogous to those found in some organelles (Borrelly et al. 2001), i.e. it carries magnesium ions from the cell in exchange for environmental protons.

The Mg^{2+}–H^+-exchanger is ATP-independent and transports magnesium ions from the cell to the environment in exchange for protons. The Mg^{2+}–H^+-exchanger operates analogously to the Na^+–H^+-exchanger (Bara et al. 1993; Borrelly et al. 2001). This ion transport mechanism is important for the regulation of intracellular pH, cellular volume and cellular growth. On the assumption that one magnesium ion is withdrawn from the cell in exchange for two protons, we obtain the following equation describing the operation of this exchanger:

$$J_{Mg-H} = C_{Mg-H}\left[n^i_{Mg}\left(n^o_H\right)^2 - n^o_{Mg}\left(n^i_H\right)^2\right] = 0, \tag{3.10}$$

where C_{Mg-H} is a constant related to the transport of magnesium ions out of the cell and the transport of protons into the cell.

The ratio of the magnesium concentrations in the cell and in the environment is calculated from formulas (3.3) and (3.10) as:

$$\frac{n^i_{Mg}}{n^o_{Mg}} = \exp\left(-\frac{2\Delta\mu_A}{3} - 2\varphi\right) = 0.036. \tag{3.11}$$

The experimental ratio of the magnesium ion concentrations on either side of the membrane is 0.04 (Smirnov et al. 2002). This good agreement between the calculated and experimental ratios supports our stoichiometric assumptions and demonstrates that this exchanger is the major mechanism of Mg^{2+} transport.

Along with positively charged ions, the cell and its environment also have anions, specifically chlorine anions, that can be carried across the membrane. In *Halobacterium* habitats, the concentrations of monovalent Cl^- and Br^- anions account for over 99.9 % of the total anion concentration (Detkova and Pusheva 2006). Transport systems for SO_4^{2-}, PO_4^{3-} and AsO_2^{1-} anions have also been found in the cells of extreme halophiles.

Chlorine's physiological significance and biological roles include the regulation of osmotic pressure in cells and tissues and the normalization of water metabolism.

However, based on the experimental ratios of chlorine ion concentrations, it can be concluded that passive transport is the major mechanism of chlorine ion transport. Chlorine ions have a Boltzmann distribution, which is represented as:

$$n^i_{Cl} = n^o_{Cl}\exp(\varphi). \tag{3.12}$$

Cells of extreme halophiles also contain non-penetrating anions (n_A) that do not have channels or transporters to carry them across the membrane. The n_A concentration in halophilic cells is approximately 6 M, given that neutral metabolites are also present (Oren 1999). In the absence of active ion transport, non-penetrating anions produce a Donnan potential.

Equating the flow of ions to zero, the condition of electroneutrality according to Melkikh and Seleznev (2009) can be written as:

$$n^o_{Cl} \exp(3\varphi) + Z_A n_A \exp(2\varphi) - \left[n^o_{Na} \exp\left(-\frac{\Delta\mu_A}{3}\right) + n^o_K \right] \cdot \exp(\varphi)$$
$$-2\left[n^o_{Ca} \exp(-\Delta\mu_A) + n^o_{Mg} \exp\left(-\frac{2\Delta\mu_A}{3}\right) \right] = 0. \quad (3.13)$$

Furthermore, by knowing the extracellular ion concentrations and the non-penetrating anion concentration ($n_A \approx 6M$), we can obtain a cubic equation for the resting potential. Because of the complexity of the analytical expression, we shall solve Eq. (3.13) numerically. The Dead Sea, a lake in the territories of Israel and Jordan, is a very typical habitat of the *Halobacterium* genus of extremely halophilic archaea. The resting potential is calculated using the ion concentrations characteristic of the Dead Sea (Table 3.1).

Substitution of these values into (3.13) gives a resting potential $\varphi_{calc.}$ of -135 mV. The calculated value of the resting potential is in good agreement with the experimental value of -125 mV (Bogomolni 1977; Bakker et al. 1976; Michel and Oesterhelt 1976).

In the following table (Table 3.2), we compare the experimentally measured intracellular ion concentrations with their calculated values (3.13) for halophilic archaea in the genus *Halobacterium*.

Thus, the calculated intracellular concentrations agree well with the corresponding experimental data.

3.3 Diatomei

In the paper (Melkikh and Bessarab 2010), the authors constructed a model of ion transport in diatom cells.

Diatoms play an important role in ecology. While accounting for almost one-third of the world's vegetation, they process approximately one-fourth of the world's carbon and transfer it from the atmosphere to the ocean. The development of a model of active ion transport in *Coscinodiscus wailesii* can provide insights into the physiology of this species and other organisms in the protist kingdom. Models for independently predicting the intracellular ion concentrations and the resting cellular potential of diatoms are currently unavailable.

Table 3.1 Extracellular ion concentrations in Halobacterium genus of archael cells

Ions	Extracellular concentrations, mM
Ca^{2+}	450
K^+	25
Na^+	1560
Cl^-	2800
Mg^{2+}	20

Table 3.2 Intracellular ion concentrations

Ions	Calculated concentrations, M	Experimental concentrations, M
Ca_i	4.5×10^{-5}	2×10^{-5}
K_i	5.5	4
Na_i	0.44	0.5
Cl_i	0.013	0.019
H_i	2.82×10^{-8}	2.6×10^{-8}
Mg_i	0.0016	0.0008

Diatoms constitute a large group of eukaryotic algae. Diatoms represent a widespread group and can be found in oceans, freshwater, and soil as well as on damp surfaces. Diatoms belong to a large group called heterokonts, which includes both autotrophs (e.g. golden algae and kelp) and heterotrophs (e.g. water molds) in the protist kingdom. There are over 10,000 known species of living diatoms. A characteristic feature of diatom cells is a unique silica (hydrated silicon dioxide) wall called a frustule. This frustule has a variety of forms, some quite beautiful and ornate, but it usually consists of two asymmetrical sides with a split between them; hence, the group name.

Diatom cells have a unique silicate (silicic acid) wall comprising two separate valves (or shells). Diatoms are traditionally divided into two orders: centric diatoms (Centrales), which are radially symmetric, and pennate diatoms (Pennales), which are bilaterally symmetric.

Diatoms are believed to play a disproportionately important role in the export of carbon from oceanic surface waters. Notably, they also play a key role in the regulation of the biogeochemical cycle of silicon in the modern ocean. They are likely responsible for 25–30 % of the world's plant production and assimilate approximately 25 % of the world's CO_2 (Boalch 1987).

Most of the diatom cell is filled with a vacuole, while the cytoplasm is arranged as a thin layer along the cellular walls. The nucleus is usually located at the center of the cell on a special cytoplasmic bridge. The cytoplasm contains lamellar or granular chloroplasts with chlorophylls, carotinoids, and xanthophylls, making the diatom yellow or yellow–brown in color. As diatoms die off, they are colored green because the chlorophyll disintegrates later than the other pigments. The auxiliary nutrients of diatoms include oils (stored in the cell as droplets), volutin, and chrysolaminarin (Ono et al. 2006).

Coscinodiscus wailesii is a centric diatom (Centrales) of the *Coscinodiscophyceae* class.

The *Coscinodiscus wailesii* cell is characterized by a membrane resting potential of approximately −90 mV (Boyd and Gradmann 1999b).

This value can change on exposure to various stimuli, such as irritants, quantitative or qualitative environmental variations, toxins, or an impaired oxygen supply.

Because sea water is the typical environment of *Coscinodiscus wailesii*, experiments with diatoms are mostly performed in solutions that are similar to ocean water. For example, in (Boyd and Gradmann 1999b) artificial sea water (isotonic to natural sea water) was prepared with the following ion concentrations: $n_{Na}^o = 0.461M$, $n_{Cl}^o = 0.520M$, $n_{Ca}^o = 0.01M$, $n_{NO_3}^o = 5 \times 10^{-6}M$, and $n_H^o = 3 \times 10^{-8}M$. Ammonium sea water solutions were prepared by substituting ammonium ions for potassium to ensure that the total ion concentration was constant $\left(n_K^o + n_{NH_4}^o = 0.01M\right)$ (Boyd and Gradmann 1999c).

The following systems of the active transport were identified in *Coscinodiscus wailesii* cells (Gradmann et al. 1993; Boyd and Gradmann 1999b, c; Bhattacharyya and Volcani 1980):

- H^+-pump (ATP-dependent proton pump);
- Na^+–NO_3^--exchanger;
- NO_3^-/Cl^--exchanger (the presence of this antiporter is assumed, but it has not been discussed in the literature because of the scarcity of experimental data);
- Ca^{+2}-pump (ATP-dependent);
- Na^+–K^+-pump (ATP-dependent);
- H^+–Cl^--exchanger;
- H^+–K^+-exchanger.

Both K^+ (Boyd and Gradmann 1999a) and NH_4^+ (Boyd and Gradmann 1999c) are transported passively.

The main transport systems in *Coscinodiscus wailesii* cells, which will be considered in this section, are shown in Fig. 3.2.

The ion concentrations inside and outside of the cell according to Boyd and Gradmann (1999b) are shown in Table 3.3.

Na^+, K^+, Ca^{+2}, H^+, and NH_4^+ are the cations that play an important role in *Coscinodiscus wailesii* cells (Boyd and Gradmann 1999b).

Sodium and potassium ions are known to be significant to any cell, including the cells of diatom *Coscinodiscus wailesii*. Sodium regulates the cell's water content and maintains its acid–base equilibrium. Potassium ions restore the

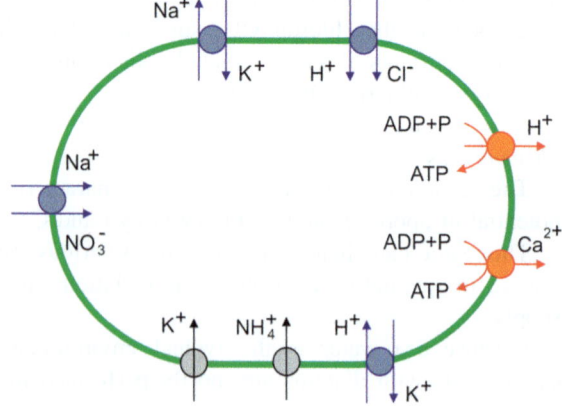

Fig. 3.2 A schematic model of ionic membrane transport in *Coscinodiscus wailesii* cells

3.3 Diatomei

Table 3.3 Ion concentrations in diatom

Ions	Intracellular concentrations, mM	Extracellular concentrations, mM
H^+	5×10^{-5}	3×10^{-5}
Ca^{2+}	1.1×10^{-5}	1
Na^+	46	461
$K^+ + NH_4$	450	10
NO_3^-	5	5×10^{-3}
Cl^-	450	520

membrane's initial potential after excitation and activate several glycolytic enzymes.

Although *Coscinodiscus wailesii* has few intracellular protons, they are also of great importance to cellular processes, including substance transport. Cells of the diatom *Coscinodiscus wailesii* are known to contain proton pumps (Gradmann et al. 1993). These pumps transfer hydrogen ions counter to electrodiffusive forces by ATP hydrolysis or other external energy sources. An ATP-dependent proton pump transfers protons out of the cell.

Let us write the proton flow produced by a H^+-ATPase in terms of the model proposed above and a "one ion—one transport system" algorithm. This flow is equal to zero as we are only considering the stationary state. We shall assume that n ions are transported at a time, resulting in the following equation:

$$J_H = C_H \left[\exp(\Delta \mu_A + n\varphi)(n_H^i)^n - (n_H^o)^n \right] = 0. \tag{3.14}$$

For a *Coscinodiscus wailesii* cell, we assume a ΔpH of -0.7, as the pH is 7.3 inside of the cell and 8 outside of the cell (Boyd and Gradmann 1999b, c).

Equating the flow (3.14) to zero yields the following expression for proton concentrations:

$$n_H^i \exp\left(\frac{\Delta \mu_A}{n} + \varphi\right) = n_H^o. \tag{3.15}$$

Using the experimental values of the potential ($\varphi = -90$ mV) and ΔpH (-0.7) at the cellular membrane and assuming that $\Delta \mu_A \approx 20$, we obtain:

$$\frac{20}{n} = \frac{90}{25} + \ln \frac{10^{-8}}{5 \times 10^{-8}} \approx 2. \tag{3.16}$$

This formula indicates that the value of n is approximately 10. However, based on experiments (Wagner et al. 2004), an H^+-ATPase that uses the hydrolysis energy of one ATP molecule, can transport a maximum of 4 protons across the membrane. Therefore, n = 4 will be used in further calculations.

The equation for the H^+-ATPase (3.14) then becomes:

$$J_H = C_H \left[\exp(\Delta \mu_A + 4\varphi)(n_H^i)^4 - (n_H^o)^4 \right] = 0. \tag{3.17}$$

The calculated ratio of proton concentrations is:

$$\frac{n_H^i}{n_H^o} = \exp\left(-\frac{\Delta\mu_A}{4} - \varphi\right) = 0.25$$

while the experimental value is:

$$\frac{n_H^i}{n_H^o} = 5.$$

Comparing these ratios, we find that the calculated and experimental values are not in agreement. Hence, it is unsatisfactory to use one proton transport system as the main system. As mentioned above, the H$^+$-ATPase is very important for maintaining intracellular pH. However, one transport system is insufficient for regulating pH. Therefore, additional mechanisms that allow protons to enter the cell need to be considered. We shall disregard the passive transport of protons, as the membrane is weakly permeable to them. Protons are known to be involved in secondary active transport, a mechanism that cells use to actively absorb or remove various substances, including ions, carbohydrates, and amino acids. The operation of secondary transport systems is directly connected to the operation of the H$^+$-ATPase. Suppose that the intracellular proton concentration is controlled by both an ATP-dependent proton pump and a secondary transport system (Briskin 1990), such as a K$^+$–H$^+$-antiporter. Let us write an equation for the ion flow produced by this exchanger, assuming that one proton enters the cell in exchange for one potassium ion:

$$J_{K-H} = C_{K-H}\left(n_K^i n_H^o - n_K^o n_H^i\right) = 0, \tag{3.18}$$

Then, the ratio of proton concentrations can be determined from the equality:

$$\frac{n_H^i}{n_H^o} = \frac{n_K^i}{n_K^o}.$$

Potassium ions are distributed passively in most cells. Suppose that this distribution is valid for the cell under examination and, therefore, that the concentration of potassium ions follows a Boltzmann distribution. Then, we obtain the following equation describing the dependence of the intracellular proton concentration on the extracellular proton concentration:

$$n_H^i = n_H^o \exp(-\varphi). \tag{3.19}$$

Thus, if the distribution of potassium ions is passive, the K$^+$–H$^+$-antiporter also produces a passive distribution of protons. Let us use Eqs. (3.17) and (3.19) to plot the dependence of intracellular pH$_i$ on extracellular pH$_o$ when only one proton transport system is in operation.

In accordance with the ideology of ion transport regulation described in Chap. 1, consider the case when two transport systems are operating simultaneously. If C in

Eq. (3.18) is assumed to be variable, but dependent on the intracellular or extracellular proton concentrations, then the intracellular proton concentration is constant when both transport systems operate simultaneously in the interval of (7.55 < pH$_o$ < 8.86) (section B—E, Fig. 3.3). The intracellular proton concentration cannot be constant beyond the interval of (7.55 < pH$_o$ < 8.86) because C cannot have a negative value.

Thus, the two transport systems (K$^+$–H$^+$-antiporter and H$^+$-ATPase), one of which has a variable capacity, can maintain a constant intracellular pH. In particular, an intracellular pH$_i$ of 7.3 will be maintained in the range of 7.55 < pH$_o$ < 8.86.

Potassium ions are among the basic intracellular ions in the diatom that are under examination. They are necessary (along with other ions) for maintaining ionic equilibrium on the inside and the outside of the cell and for stabilizing enzymes, membranes, and other cellular structures.

Passive transport is most likely the main transport system for potassium ions in diatoms. The *Coscinodiscus wailesii* cell membrane is highly permeable to potassium ions (Boyd and Gradmann 1999b). The concentration of potassium ions is assumed to follow a Boltzmann distribution. In other words, the concentration can be expressed as:

$$n_K^i = n_K^o \exp(-\varphi).$$

Comparing the following calculated ratio

$$\frac{n_K^i}{n_K^o} = \exp(-\varphi) = 36.6,$$

to the following experimental value:

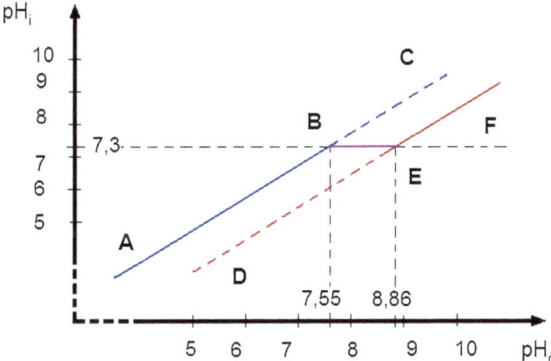

Fig. 3.3 The dependence of the intracellular pH$_i$ on the extracellular pH$_o$ (Melkikh and Bessarab 2010). The straight line ABC is the pH$_o$ dependence of the intracellular pH$_i$ when protons are carried by a K$^+$–H$^+$-antiporter. The straight line DEF is the pH$_o$ dependence of the intracellular pH$_i$ when protons are carried by an H$^+$-ATPase

$$\frac{n_K^i}{n_K^o} = 45,$$

we see that passive transport is the main transport system for potassium ions.

According to Bhattacharyya and Volcani (1980), the sodium ions of *Coscinodiscus wailesii* cells are transported by a Na$^+$-K$^+$-ATPase that contains three sorption centers for sodium ions and two sorption centers for potassium ions. Assuming that the active flow is proportional to the third power of the sodium ion concentration and the second power of the potassium ion concentration, we obtain an equation for the active flow of sodium ions (C_{Na-K} is a constant that includes parameters related to the ion and the biomembrane):

$$J_{Na-K} = C_{Na-K}\left[\exp(\Delta\mu_A + \varphi)(n_{Na}^i)^3(n_K^o)^2 - (n_{Na}^o)^3(n_K^i)^2\right]. \qquad (3.20)$$

Ions have an asymmetrical effect on one another when they are carried by different transport systems. An example of this asymmetry is that potassium ions make a considerable contribution to the flow of protons, while protons make a relatively small contribution to the transport of potassium ions (it was concluded from Eq. (3.8) that most potassium ions are carried passively). The same can be said about sodium ions; namely, the distribution of potassium ions has a strong effect on sodium ions (according to formula (2.1)), whereas the transport of sodium ions has a weak effect on potassium ions.

Let us calculate the ratio of the intracellular and extracellular concentrations of sodium ions. When three sodium ions and two potassium ions are transported, we have:

$$\frac{n_{Na}^i}{n_{Na}^o} = \sqrt[3]{\left(\frac{n_K^i}{n_K^o}\right)^2 \cdot \exp(-\Delta\mu_A - \varphi)} \approx 0.05 \qquad (3.21)$$

The experimental value of the ratio between the intracellular and extracellular concentrations of sodium ions is (Boyd and Gradmann 1999c):

$$\frac{n_{Na}^i}{n_{Na}^o} = 0.1.$$

Therefore, this pump is the main system for transporting sodium ions. The passive flow of sodium ions can be disregarded, as sodium is toxic to the cell. Furthermore, the main transport system will be the pump system, which removes sodium from the cell. The calculated ratio of the intracellular and extracellular sodium ion concentrations is in good agreement with the experimental ratio, confirming that we have chosen the correct Na$^+$-K$^+$-ATPase pump stoichiometry.

The active flow of calcium ions is due to Ca^{+2}-ATPase, which removes calcium ions from the cell. Calcium is responsible for the structural and functional intactness of the membrane and participates in the cell's signal processes.

3.3 Diatomei

The Ca^{+2}-ATPase performs active transport of calcium ions across the cellular membrane and maintains a intracellular concentration of these ions lower than their extracellular concentration (Gradmann and Boyd 2000). The pump, which is located in the cytoplasmic membrane, binds calcium ions and carries them out of the cell at the expense of ATP energy.

Assuming that one calcium ion is actively transported out of the cell, the mathematical expression for the operation of the Ca^{+2}-ATPase is written in the form:

$$J_{Ca} = C_{Ca}\left[\exp(\Delta\mu_A + 2\varphi)n^i_{Ca} - n^o_{Ca}\right] = 0, \tag{3.22}$$

where C_{Ca} is a constant related to the active transport of calcium ions.

The calculated ratio of the intracellular and extracellular concentrations of calcium ions is:

$$\frac{n^i_{Ca}}{n^o_{Ca}} = \exp(-\Delta\mu - 2\varphi) \approx 2.8 \times 10^{-5}.$$

This value is in good agreement with the corresponding experimental ratio (Brownlee et al. 1987):

$$\frac{n^i_{Ca}}{n^o_{Ca}} = 1.1 \times 10^{-4}.$$

Therefore, we shall neglect the passive flow of these ions.

According to Boyd and Gradmann (1999c), the transport of NH_4^+ ions across the membrane is mostly passive. Then, in line with the Boltzmann distribution, we can write:

$$n^i_{NH_4} = n^o_{NH_4} \exp(-\varphi). \tag{3.23}$$

To ascertain the passive distribution of ammonium in the *Coscinodiscus wailesii* cell, we calculate the ratio of the intracellular and extracellular concentrations of ammonium ions as:

$$\frac{n^i_{NH_4}}{n^o_{NH_4}} = \exp(-\varphi) \approx 37. \tag{3.24}$$

The experimental ratio of the intracellular and extracellular concentrations of ammonium ions is 45 (Boyd and Gradmann 1999c), confirming the passive distribution of ammonium ions.

Along with cations, the cell and its environment also contain anions. The transport systems of Cl^- and NO_3^- ions are known.

Chlorine ions are physiologically and biologically significant as they control the osmotic pressure of cells and tissues and play an important role in the normalization of water exchange. Chlorine is among the most abundant and important elements in a cell.

Passive transport is also assumed for chlorine ions, i.e. they have a Boltzmann distribution:

$$n_{Cl}^i = n_{Cl}^o \exp(\varphi). \qquad (3.25)$$

The intracellular concentration of chlorine ions ($n_{Cl}^i \approx 14\,\text{mM}$) can be found from Eq. (3.25). The experimental value of the intracellular concentration of chlorine ions is $n_{Cl}^i = 450\,\text{mM}$ (Boyd and Gradmann 1999c). A comparison of the calculated and experimental values suggests that chlorine ions should have an active transport system in the cells under examination.

According to Melkikh and Bessarab (2010), we assume that an H^+–Cl^--symporter carries chlorine ions in *Coscinodiscus wailesii* cells. This assumption follows from the presence of an H^+–Cl^--symporter in many plant cells with features similar to those of *Coscinodiscus wailesii* cells (Gradmann et al. 1993). We shall assume that 1 chlorine ion and 2 protons are carried simultaneously:

$$J_{H-Cl} = C_{H-Cl}\left((n_H^i)^2 \cdot n_{Cl}^i \exp(\varphi) - (n_H^o)^2 \cdot n_{Cl}^o\right) = 0. \qquad (3.26)$$

The resulting calculated value is:

$$\frac{n_{Cl}^i}{n_{Cl}^o} = \left(\frac{n_H^o}{n_H^i}\right)^2 \cdot \exp(-\varphi) = 1.2. \qquad (3.27)$$

The experimental value of the concentration ratio of chlorine ions on the inside and the outside of the cell is 0.86.

Let us turn to the transport of nitrate ions across the membrane of *Coscinodiscus wailesii* cells. Similar to Boyd and Gradmann (1999c), we assume that the Na^+–NO_3^--symporter is the transport system for nitrate ions. Assuming that n sodium ions and one nitrate ion are carried simultaneously, we have the equation:

$$J_{Na-NO_3} = C_{Na-NO_3}\left((n_{Na}^i)^n \cdot n_{NO_3}^i \cdot \exp((n-1)\varphi) - (n_{Na}^o)^n \cdot n_{NO_3}^o\right) = 0. \qquad (3.28)$$

Considering the available concentration ratios, we let $n = 2$ and calculate the ratio of intracellular and extracellular concentrations of nitrate ions from formula (3.28). The calculated value is:

$$n_{NO_3}^i = \left(\frac{n_{Na}^o}{n_{Na}^i}\right)^2 \cdot n_{NO_3}^o \cdot \exp(-\varphi) = 7.3\,\text{mM}. \qquad (3.29)$$

This value agrees well with the experimental value ($n_{NO_3}^i = 5\,\text{mM}$) (Boyd and Gradmann 1999c), confirming that we have correctly chosen the stoichiometry of the Na^+–NO_3^--exchanger.

Therefore, the equation for the flow of nitrate ions has the form:

3.3 Diatomei

$$J_{Na-NO_3} = C_{Na-NO_3}\left((n^i_{Na})^2 \cdot n^i_{NO_3} \cdot \exp(\varphi) - (n^o_{Na})^2 \cdot n^o_{NO_3}\right) = 0. \quad (3.30)$$

This set of equations should be supplemented with the condition of electroneutrality in the *Coscinodiscus wailesii* cell:

$$n^i_{Na} + n^i_H + n^i_K + 2n^i_{Ca} + n^i_{NH_4} = n^i_{Cl} + n^i_{NO_3} + Z_A n_A. \quad (3.31)$$

Substituting the concentrations of basic ions in expression (3.31) and using $Z_A = 1$, it is possible to calculate the concentration of non-penetrating ions in the cell ($n_A = 63.5$ mM).

For the electroneutrality condition, we have:

$$n^o_{Na}\exp\left(-\frac{\Delta\mu_A}{3} - \varphi\right) + n^o_{K+NH_4}\exp(-\varphi) = n^o_{Cl}\exp\left(\varphi + \frac{\Delta\mu_A}{6}\right) + Z_A n_A. \quad (3.32)$$

The solution for φ takes the form:

$$\varphi = \ln\left(\sqrt{\left(\frac{Z_A n_A}{2n^o_{Cl}\exp\left(\frac{\Delta\mu_A}{6}\right)}\right)^2 + \left(\frac{n^o_{Na}}{n^o_{Cl}\exp\left(\frac{\Delta\mu_A}{6}\right)}\right)\exp\left(-\frac{\Delta\mu_A}{3}\right) + \frac{n^o_K}{n^o_{Cl}\exp\left(\frac{\Delta\mu_A}{6}\right)}}\right.$$
$$\left. - \frac{Z_A n_A}{2n^o_{Cl}\exp\left(\frac{\Delta\mu_A}{6}\right)}\right). \quad (3.33)$$

Substituting the ion concentrations in artificial sea water and the non-penetrating anion concentration in Eq. (3.33) gives a resting potential of $\varphi_{calc.} \approx -92.5$ mV. The calculated value of the resting potential is in good agreement with its experimental value [−90 mV (Boyd and Gradmann, 1999a, b, c)].

The calculated intracellular ion concentrations in the *Coscinodiscus wailesii* diatom are compared to the corresponding experimental values in Table 3.4.

It is seen from Table 3.4 that the intracellular ion concentrations and the resting potential are in good qualitative agreement.

In many situations, when environmental conditions change, it is necessary to know what corresponding changes will occur in specific ion concentrations. This issue is especially significant with respect to changes in environmental salinity, which leads to changes in the concentrations of NaCl, KCl, and other salts.

Let us plot the resting potential at the *Coscinodiscus wailesii* cellular membrane as a function of the potassium ion concentration, assuming that K is added with Cl because KCl is supplied in the environment.

The addition to the KCl concentration is labeled as α. Then, Eq. (3.33) for the resting potential takes the form:

Table 3.4 Ion concentrations

Ions	Calculated intracellular concentrations, mM	Experimental intracellular concentrations, mM
H^+	5×10^{-5}	5×10^{-5}
Ca^{2+}	2.8×10^{-5}	1.1×10^{-5}
Na^+	21.5	46
$K^+ + NH_4^+$	366	450
NO_3^-	7.3	5
Cl^-	650	450

$$\varphi = \ln\left(\sqrt{\left(\frac{Z_A n_A}{2(n_{Cl}^o + \alpha)\exp\left(\frac{\Delta\mu_A}{6}\right)}\right)^2 + \frac{n_{Na}^o}{(n_{Cl}^o + \alpha)\exp\left(\frac{\Delta\mu_A}{6}\right)}\exp\left(-\frac{\Delta\mu_A}{3}\right) + \frac{(n_K^o + \alpha)}{(n_{Cl}^o + \alpha)\exp\left(\frac{\Delta\mu_A}{6}\right)}} - \frac{Z_A n_A}{2(n_{Cl}^o + \alpha)\exp\left(\frac{\Delta\mu_A}{6}\right)}\right).$$

(3.34)

Figure 3.4 shows the experimental (Boyd and Gradmann 1999b) and calculated values of the resting potential at the *Coscinodiscus wailesii* cellular membrane (in relative units) based on the extracellular potassium ion concentration (in mM). From this figure, we can see that the potential increases (in its absolute value) with an increase in the potassium ion concentration.

Figure 3.4 shows that the values of the resting potential agree with the corresponding experimental data (Boyd and Gradmann 1999b). Therefore, we can conclude that the calculations are correct. Notice that the proposed model does not include regulatory mechanisms for the resting membrane potential or intracellular ion concentrations, which are depending on the environmental composition (only one main transport system is assumed for each ion). This means that variation in resting potential can be predicted with sufficient accuracy over a relatively narrow interval as of extracellular ion concentrations before regulatory mechanisms come into play.

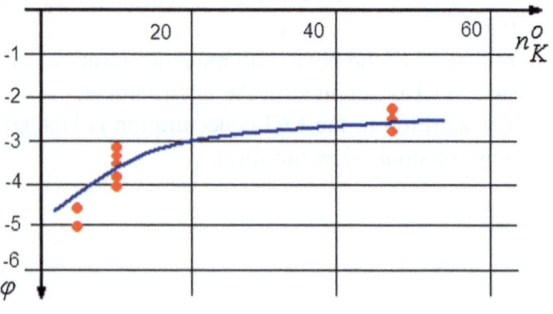

Fig. 3.4 The dependence of the resting potential on the extracellular potassium ion concentration according to (Melkikh and Bessarab 2010)

3.4 E. coli

E. coli is one of the most intensively studied microorganisms. In the past decades, substantial progress has been made in studying the composition and regulation of the *E. coli* genome as well as its metabolism and other processes. Systems biology is a promising direction for the investigation of cells, including *E. coli*. This direction is based on a systemic approach, modeling cells (organisms) as a system. However, the system of active ion transport in *E. coli*, which is one of the most important subsystems in the cell, is not clearly understood. Models of active ion transport in *E. coli* are especially scarce. Because models of active ion transport are unavailable, it is difficult to fully understand the processes taking place in *E. coli*.

3.4.1 A Transport Model of Basic Ions in E. coli

Colibacillus is a symbiotic gram-negative bacterium. It is shaped like an oblong rod with cup points (0.4–0.8 × 1–3 μm). It is mobile in natural conditions, is a facultative anaerobe, and ferments glucose, lactose and other hydrocarbons. *E. coli* is among the most typical representatives of normal gut organisms in mammals. However, they can reside in other media as well. *E. coli*'s transport subsystem carries nutrients into the cell, maintains its basic ion concentrations and its resting potential at the required levels, and controls its osmotic pressure.

E. coli's biological membrane has several types of ionic pumps, which are molecular protein structures that are built into the membrane and are responsible for the active transport of substances towards their higher electrochemical potential. The ionic pumps are driven by the free energy of the ATP hydrolysis.

Along with the ionic pumps, many ions are carried by exchangers using the free energy of other ions. In this case, two or more types of ions are transported simultaneously across the membrane. Passive transport across the membrane is also available for each type of ion.

The resting potential of the *E. coli* membrane is approximately −110 mV (Eisenbach 1982; Padan et al. 2005; Grogan 2001). Maintaining a constant potential is important for cells because many ion transport processes (both active and passive forms) considerably depend on the resting potential.

According to data in the literature, an *E. coli* cell has the following active transport systems:

- H^+-ATPase;
- Na^+-H^+-exchanger, which is independent of ATP;
- Ca^{2+}-ATPase;
- H^+–K^+-pump (ATP-dependent)

Magnesium ions may also be carried by an active transport system (this assumption will be maintained below), but data on the magnesium transport system are contradictory.

Many ions are also carried across the membrane by passive transport.

These processes are shown schematically in Fig. 3.5.

Sodium, potassium, calcium and magnesium ions have the largest cation concentrations in the internal environment of *E. coli*. The cell also contains ions of iron, manganese, zinc, copper, molybdenum and other important microelements. Microelement transport presents a separate problem and is not discussed herein. This approximation is taken because there is a small concentration of free microelements in both the cell and its environment. This factor also results in their small contribution to the resting potential.

Even though the cellular proton concentration is small (approximately 10^{-7} M), protons play a very important role in transport systems because they are used by exchangers.

Chlorine ions and non-penetrating negative ions are the main anions in the cell. Concentrations of other anions (e.g. amino acids) are relatively small and can be neglected in the total ionic balance.

We shall simulate ion flow in terms of a model proposed earlier. ATP production in *E. coli* is not discussed here because this system should be considered separately.

In accordance with the model proposed earlier, the flow of ions produced by an H$^+$-ATPase is written as:

$$J_H = C_H \left[\exp(\Delta \mu_A + m\varphi)\left(n_H^i\right)^m - \left(n_H^o\right)^m \right] = 0, \qquad (3.35)$$

where C_H is the proton transport constant, which is proportional to the H$^+$-ATPase's operating frequency and the number of protons in the cell.

The parameter m characterizes stoichiometric properties of an H$^+$-ATPase and has the following limits: if the pH difference (ΔpH) at the membrane is relatively large (3–4 units), they transport two protons per ATP molecule; if the pH difference is relatively small (1–2 units), they carry four protons per ATP molecule.

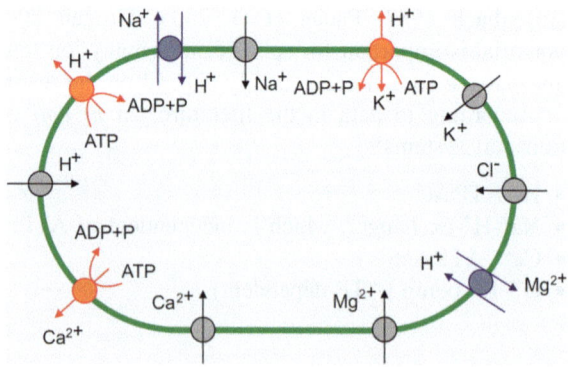

Fig. 3.5 Transport processes in *E. coli*

3.4 E. coli

These properties are also characteristic of other V–H$^+$-ATPases, including the H$^+$-ATPase of *E. coli*.

According to experimental data, the typical ΔpH value for *E. coli* is approximately 0.7–0.9 (Padan et al. 1976, 2005; Grogan 2001). In other words, it may be assumed that the colibacillus cell's H$^+$-ATPase carries 3–4 protons across its membrane.

We shall consider the cell in its stationary state. Therefore, the total flow of each ion type should be zero. Neglecting the passive transport of protons across the membrane and equating (3.35) to zero, we have:

$$n_H^i \exp\left(\frac{\Delta\mu_A}{m} + \varphi\right) = n_H^o.$$

This equation can be rearranged to the form:

$$\frac{\Delta\mu_A}{m} = -\varphi + \ln\frac{n_H^o}{n_H^i} = \Delta\mu_H,$$

where $\Delta\mu_H$ is the difference of electrochemical potentials of protons on both sides of the membrane.

Using experimental values of the potential ($\varphi = -110$ mV) and the ΔpH across the cellular membrane (0.8) along with a value of $\Delta\mu_A \approx 20$, we obtain:

$$\frac{20}{m} = \frac{110}{25} + 2.3 \times 0.8 \approx 6.$$

It can be seen from this formula that the value of m is between 3 and 4.

We can express the proton concentration of the cell as:

$$n_H^i = n_H^o \exp\left(-\frac{\Delta\mu_A}{m} - \varphi\right). \tag{3.36}$$

Let us consider the transport of sodium ions by the Na$^+$–H$^+$ exchanger. This exchanger pumps sodium ions out of the cell.

To write an equation for the flow of ions in the Na$^+$–H$^+$ exchanger, we assume that one sodium ion is released from the cell in exchange for one proton:

$$J_{Na-H} = C_{Na-H}(n_{Na}^i n_H^o - n_{Na}^o n_H^i) = 0, \tag{3.37}$$

where C_{Na-H} is a constant related to the transport of sodium ions out of the cell and the transport of protons into the cell. This constant is proportional to the operating frequency of the exchanger and the number of the exchangers in the cell.

By disregarding the passive transport of sodium ions across the membrane, equating the flow of sodium ions (3.37) to zero and using the formula for the proton concentration at $m = 3$, we have:

$$\frac{n_{Na}^i}{n_{Na}^o} = \exp\left(-\frac{\Delta\mu_A}{3} - \varphi\right).$$

The number of potassium ions in the environment is much smaller than its number inside the colibacillus cell. Although *E. coli* cells have an H^+–K^+ pump that is ATP-dependent and exchanges cellular protons for environmental potassium ions, the distribution of potassium ions approaches a passive distribution. We shall therefore assume that potassium ions obey a Boltzmann distribution:

$$n_K^i = n_K^o \exp(-\varphi).$$

The role of the H^+-K^+ pump will be discussed separately below.

Let us consider the transport of Ca^{2+} ions. Specifically, calcium participates in cellular signaling processes. To realize this function, the cell's calcium concentration must be maintained at a low level. The active transport of calcium ions is due to a Ca^{2+} pump that removes calcium ions from the cell.

The equation for calcium ions only accounts for the operation of the ATP-dependent pump and disregards the passive transport of these ions. The equation for the flow of calcium ions can then be written as follows, assuming that one calcium ion is actively removed from the cell:

$$J_{Ca} = C_{Ca}\left[\exp(\Delta\mu_A + 2\varphi)n_{Ca}^i - n_{Ca}^o\right] = 0. \tag{3.38}$$

where C_{Ca} is a constant related to the active transport of calcium.

The deduced equation reflects the stoichiometry of the Ca^{2+}-ATPase pump: one calcium ion is carried across the membrane to the environment during hydrolysis of one ATP molecule.

Equating the flow (3.38) to zero, we obtain the concentration of calcium ions in the cell:

$$n_{Ca}^i = n_{Ca}^o \exp(-\Delta\mu_A - 2\varphi).$$

Magnesium is a universal regulator of the physiological and biochemical processes that occur in cells. Magnesium ions control the operation of ATP-dependent exchangers. Magnesium participates in regulating the concentration and transport of calcium, potassium, sodium and phosphate ions both inside and outside of the cell. Magnesium interacts with cellular lipids and ensures the intactness of the cellular membrane.

The literature contains rather contradictory data on the transport of Mg^{2+} ions. Specifically, some facts point to the existence of a Mg^{2+}-ATPase. Let us consider possible transport variants of Mg^{2+} ions and compare the results with the corresponding experimental data. First, let us assume that Mg^{2+} is distributed passively. In this case, we can write an equation that is analogous to the equation for potassium ions:

$$n_{Mg}^i = n_{Mg}^o \exp(-2\varphi). \tag{3.39}$$

Substituting the known value of *E. coli*'s resting potential (-110 mV or $\varphi \approx -4$ in dimensionless units) into formula (3.39) we have a concentration ratio of approximately 300 for magnesium ions inside and outside of the cell. This result

3.4 E. coli

differs from the actual value by several orders of magnitude. According to experimental data, approximately 1 and 10 mM of Mg^{2+} are present in the external and internal environments, respectively. However, note that the cellular concentration of Mg^{2+} ions can be lower than these values (e.g. most of the magnesium is bound, so some of these ions are not in the solution). The derived flow equations only account for ions in the solution.

Moreover, if Mg^{2+} ions had a passive distribution, the cell's resting potential would be much smaller than -100 mV.

Let us assume that Mg^{2+} ions are transported in manner similar to calcium transport, that is, by means of a Mg^{2+}-ATPase. Then, the intracellular concentration is given by an equation analogous to formula (3.39):

$$n^i_{Mg} = n^o_{Mg} \exp(-\Delta\mu_A - 2\varphi).$$

Substituting the known values of $\Delta\mu_A \approx 20$ and $\varphi \approx -4$, the concentration ratio of magnesium outside and inside the cell is nearly 1.6×10^5 (i.e. magnesium is practically absent from the cell). This result also differs from the measured value by several orders of magnitude. Based on the obtained values, we can conclude that E. coli must have another active transport system for magnesium. For example, a Mg^{2+}–H^+ exchanger has been found in some of the simplest cells and organelles. Let us assume that this exchanger, which transports one magnesium ion in exchange for one proton, is present in E. coli. We can write an equation for the flow of magnesium ions:

$$J_{Mg-H} = C_{Mg-H}\left[n^i_{Mg}\left(n^o_H\right)\exp(\varphi) - n^o_{Mg}\left(n^i_H\right)\right], \quad (3.40)$$

where C_{Mg-H} is a constant that is proportional to the operating frequency of the exchanger and the number of the exchangers in the cell. We assume that magnesium's other active transport systems make a smaller contribution to its flow. By equating the flow (3.40) to zero and substituting (3.36), the concentration of magnesium ions in the cell is:

$$n^i_{Mg} = n^o_{Mg} \exp\left(-\frac{\Delta\mu_A}{m} - 2\varphi\right).$$

Substituting the corresponding numerical values, we obtain a concentration ratio of magnesium on the inside and the outside of the cell equal to 20 at $m = 4$ and 3.8 at $m = 3$. The latter is closer to the experimental value. It can be shown that there is comparatively less agreement with the experimental data for other stoichiometries of magnesium and proton transport.

Let us assume that chlorine ions, similar to potassium ions, are distributed by Boltzmann's law. Then, we can write the following:

$$n^i_{Cl} = n^o_{Cl} \exp(\varphi).$$

Substituting the known value of the resting potential, we have a concentration ratio of chlorine ions on the outside and the inside of the cell of approximately 50, which approaches the experimental value.

To theoretically determine the resting potential of E. coli, we shall consider the condition for electroneutrality of the medium in the cell:

$$n^i_{Na} + n^i_H + n^i_K + 2n^i_{Ca} + 2n^i_{Mg} = n^i_{Cl} + Z_A n_A.$$

The value of $Z_A n_A$ can be calculated from experimentally determined concentrations of dissimilar ions in the cell. Calculations for E. coli give $Z_A n_A \approx 200$ mM.

Let the electroneutrality condition only hold for ions with concentrations that are sufficiently large within the cell:

$$n^o_K \exp(-\varphi) + n^o_{Na} \exp\left(-\frac{\Delta\mu_A}{3} - \varphi\right) = n^o_{Cl} \exp(\varphi) + Z_A n_A. \qquad (3.41)$$

Solving (3.41) for the potential gives an expression for φ:

$$\varphi = \ln\left(\sqrt{\left(\frac{Z_A n_A}{2n^o_{Cl}}\right)^2 + \left(\frac{n^o_{Na}}{n^o_{Cl}} \exp\left(-\frac{\Delta\mu_A}{3}\right) + \frac{n^o_K}{n^o_{Cl}}\right)} - \frac{Z_A n_A}{2n^o_{Cl}}\right). \qquad (3.42)$$

If the ion concentrations outside the cell are known, it is possible to calculate the resting potential. Known external concentrations of the major ions for E. coli are shown in Table 3.5. (Garrett and Grisham 2002; Neidhardt 1987; Nanninga 1985; Ingraham et al. 1983).

The environmental neutrality in Table 3.5 is maintained by other anions (e.g. HCO_3^- anion) that are not considered in this model.

Substituting these values into (3.42), we obtain a resting potential approximately equal to $\varphi \approx -92$ mV (3.68 in dimensionless units). This value is in good agreement with the theoretical resting potential (approximately -110 mV). Substituting the obtained resting potential into the ion concentration formulas, we obtain the values tabulated in Table 3.6. (Garrett and Grisham 2002; Neidhardt 1987; Nanninga 1985; Ingraham et al. 1983).

Table 3.6 shows that the theoretical and experimental values are in good qualitative agreement.

Table 3.5 Known external concentrations of ions for E. coli

Ion	External concentration, mM
Ca^{2+}	2.5
K^+	5.6
Na^+	129.6
Cl^-	99.5
H^+	3.2×10^{-4}
Mg^{2+}	0.85

3.4 E. coli

Table 3.6 Theoretical and experimental ion concentrations for *E. coli*

Ion	Calculated concentration, mM	Experimental concentration, mM
Ca^{2+}	8×10^{-6}	1×10^{-4}
K^+	199	200
Na^+	7.2	5
Cl^-	2.8	6
H^+	1.6×10^{-5}	4.4×10^{-5}
Mg^{2+}	1.7	10 (bound incl.)
Resting potential, mV	−92	−110

We derive the dependence of the membrane potential on the extracellular concentrations of major ion types without regulation.

Let us establish the dependences of the intracellular ion concentrations and the resting potential on changes in environmental potassium ion concentrations. We shall assume that potassium is added simultaneously with chlorine. The addition of these concentrations will be labeled as α. Then, from (3.42) we have:

$$\varphi = \ln\left(\sqrt{\left(\frac{Z_A n_A}{2n_{Cl}^o + 2\alpha}\right)^2 + \left(\frac{n_{Na}^o}{n_{Cl}^o + \alpha}\exp\left(-\frac{\Delta\mu_A}{3}\right) + \frac{n_K^o + \alpha}{n_{Cl}^o + \alpha}\right)} - \frac{Z_A n_A}{2n_{Cl}^o + 2\alpha}\right). \tag{3.43}$$

Figure 3.6 presents the dependence of the resting potential of the colibacillus cell on an addition to the extracellular potassium ion concentration α. This figure shows that the negative potential decreases in absolute value as the environmental potassium concentration increases. The reason for this behavior is that, as the environmental potassium and chlorine concentrations rise, the fraction of the passively transported ions increases relative to the fraction of actively transported ions (sodium).

If we neglect the contribution of the terms n_{Na}^o and n_{Cl}^o in (3.43), we have:

$$\varphi = \ln\left(\frac{n_K^o}{Z_A n_A}\right).$$

This simplified dependence (dashed line) is shown in Fig. 3.6 for comparison with the exact result (red line).

Let us next consider the dependence of the intracellular ion concentrations of *E. coli* on changes in environmental potassium concentrations. The concentrations of magnesium, calcium, sodium and protons are defined by the factor $\exp(-\varphi)$. Therefore, these concentrations are approximately proportional to $\frac{n_K^o}{Z_A n_A}$.

The intracellular potassium concentration is nearly independent of the external concentration, while the intracellular chlorine concentration is proportional to its external concentration.

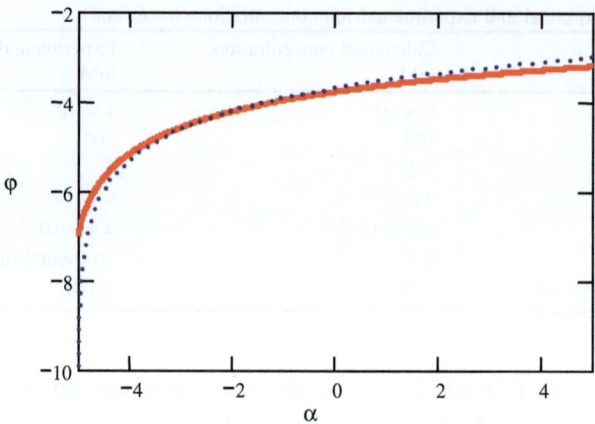

Fig. 3.6 The dependence of the resting potential on an addition to the environmental potassium ion concentration

Because the proton concentration is small (protons are not potential-forming ions), the intracellular proton concentration has a linear dependence on its extracellular concentration, as follows from formula (3.36). The same applies to all ions with relatively small concentrations.

If the sodium ion concentration changes in the environment, we obtain the following formula for the potential:

$$\varphi = \ln\left(\sqrt{\left(\frac{Z_A n_A}{2n_0^{Cl} + 2\alpha}\right)^2 + \left(\frac{(n_o^{Na} + \alpha)}{n_0^{Cl} + \alpha}\right)\exp\left(-\frac{\Delta\mu_A}{3}\right) + \frac{n_o^K}{n_0^{Cl} + \alpha}} - \frac{Z_A n_A}{2n_0^{Cl} + 2\alpha}\right). \tag{3.44}$$

Here, α is an addition to the environmental sodium and chlorine concentrations. Based on formula (3.44) we can conclude that the resting potential weakly depends on the environmental concentration of sodium ions (less than 1 % as the concentration of sodium ions doubles). This is because, on the one hand, the fraction of active ions increases as sodium is added (leading to an increase in the absolute value of the potential) while, on the other hand, the fraction of non-penetrating ions decreases with respect to the number of chlorine ions (leading to a decrease in the absolute value of the potential). Thus, these two effects essentially compensate for one another. Another important fact is that the cellular sodium ion concentration is low, making small contributions to the electroneutrality condition and, consequently, to the resting potential. Hence, it follows that all other ion concentrations (except for chlorine) weakly depend on the environmental sodium concentration. Not that we only consider relatively small deviations in ion concentrations here. Additional control mechanisms are typically engaged if the deviations are large.

3.4.2 Calculation of the Osmotic Pressure Differences in Bacteria

It is possible to calculate the osmotic pressure from the ion concentrations. It is important for cells to maintain their osmotic pressure because their biomembrane can be damaged if the cellular pressure is too high. To avoid rupture, many bacteria have a wall that withstands pressures up to dozens of atmospheres.

Let us consider two cases: *E. coli* and archaeabacteria. In the first case, we can use formula (1.3) for a first approximation:

$$p = kT \sum_i n_i.$$

Substituting all the parameters, we obtain an environmental osmotic pressure equal to approximately $P^o \approx 6.3$ atm and an internal pressure equal to approximately $P^i \approx 10$ atm.

As a result, we can determine the pressure difference:

$$\Delta P = P^i - P^o = 3.7 \, \text{atm}.$$

Let us use the obtained intracellular and extracellular concentrations to calculate the pressure difference across the membrane of archaeabacteria. Note that the solution is very strong on the inside and the outside of the cell. Therefore, the pressure of the solution cannot be calculated from formulas such as $p = nkT$, which are only valid for dilute solutions. In accordance with (1.5), we shall write the pressure as:

$$p = kT \sum_i n_i \gamma_i,$$

where:

$$\lg \gamma_i = -0.509 z_i^2 \left(\frac{1}{2} \sum_i z_i^2 m_i \right)^{1/2}$$

with m_i representing the number of moles in 1000 g of solvent.

Neglecting the calcium ions, we have the following result for the intracellular medium (for charged ions only):

$$p^i = RT \sum_i C_i \gamma_i \approx 1.8 \times 10^6 \, Pa.$$

The equation for the extracellular medium is:

$$p^i = RT \sum_i C_i \gamma_i \approx 1.4 \times 10^6 \, Pa.$$

The pressure difference on the outside and the inside of the cell is approximately 4 atm. Considering that archaeabacteria have a cell wall and are capable of withstanding considerable pressure differentials, the obtained value is relatively small.

3.5 Regulation of Ion Transport in Select Microorganisms

Microorganisms use a number of strategies in response to changes in environmental conditions. For example, two possible strategies can be enacted in response to salt stress (see, for example, Oren 1999; Ventosa et al. 1998):

1. The salt-in strategy is used when cells can support high intracellular salt concentrations allowing them to conform to the high environmental salt concentrations.
2. Cells can manufacture their own osmolites and implement regulatory mechanisms to transport of ions.

Extremely halophilic archaeabacteria of the *Halobacterium* order adhere to the first strategy. For cells using this adaptive strategy, all enzymatic and structural components should function in the presence of a high internal salt concentration.

In the majority of the other halophilic and halotolerant organisms, osmotic balance is achieved by the synthesis of small organic molecules or by absorption of these molecules from the environment. In yeast cells, only the strategy of substituted soluble substances is observed.

Cells use three strategies to maintain an optimal potassium concentration and a stable relative internal concentration of K^+/Na^+: (1) strict discrimination among the influx of alkali metal cations (transporters have a higher affinity for potassium than for sodium), (2) efficient efflux of toxic cations from the cell, and (3) selective sequestration (compartmentalization) of cations in organelles (Nass and Cunningham 1997; Serrano and Rodriguez-Navarro 2001; Kinclova et al. 2002; Sychrova 2004).

According to experimental data, additional mechanisms switch on in response to changes in external ion concentrations to maintain the required levels of intracellular ion concentrations. These mechanisms can work simultaneously with basic transport system, or the cell can switch from one system to another.

Let us consider two examples. First, we will discuss the bacterial regulation of potassium ion transport when a small amount of potassium ions is present in the environment. Then, we will consider yeast's regulation of sodium ion transport during salt stress.

Bacteria and yeast differ from other organisms because of their ability to grow in a very wide range of environmental ion concentrations. This is specifically achieved by switching their auxiliary (regulatory) ion transport systems.

Let us consider the regulatory model potassium ion transport in bacteria.

As a rule, the environmental potassium ion concentration is much lower than its concentration inside a cell. In the majority of microorganisms, potassium ions are passively transported (based on their potential), though they also participate in a

3.5 Regulation of Ion Transport in Select Microorganisms

number of transport systems. However, this passive mechanism cannot provide a high enough potassium ion concentration inside the cell, if the environmental potassium concentration drops by an order of magnitude. Experimental data show that the H^+–K^+-pump becomes the most essential potassium carrier in this situation. The pump depends on ATP and exchanges cellular protons for environmental potassium ions. After exchanging one potassium ion for two protons across the membrane, a net result of one positive charge is transported outside the cell.

The equation describing the work of proton-potassium pump can be written according to the model above:

$$J_{H-K} = C_{H-K}\left(\exp(\Delta\mu_A + \varphi)(n_H^i)^2 n_K^o - (n_H^o)^2 n_K^i\right),$$

where C_{H-K} is a constant related to the transport of potassium ions into the cell and the transport of protons out of a cell.

The equation for the proton flux, which is created by an H^+-ATPase, can be written as follows:

$$J_H = C_H\left[\exp(\Delta\mu_A + 2\varphi)(n_H^i)^2 - (n_H^o)^2\right] = 0,$$

where C_H is a constant related to the transport of protons out of a cell.

Equating fluxes to zero, we obtain:

$$n_K^i = n_K^o \exp\left(\frac{\Delta\mu_A}{3} - \varphi\right) \quad (3.45)$$

To determine an optimum strategy of ion transport when potassium is lacking, we adopt passive transport as a secondary strategy. The ratio between the potassium concentrations is the following:

$$n_K^i = n_K^o \exp(-\varphi). \quad (3.46)$$

Let us construct dependences (3.45) and (3.46) while taking into account that the potential will also depend on the potassium ion concentration. However, at small environmental potassium ion concentrations (including at aspiration, when its concentration is zero), the resting potential approaches a constant (see, for example, Eq. 3.42). Therefore, we shall neglect any changes in the potential for this case. The specified dependences are shown in Fig. 3.7.

Using the effective strategy (in which only one transport system is active at any given moment), it is possible to conclude that switching from one transport system to another should occur when the external potassium concentration is approximately equal to 0.1 mM (Fig. 3.7). This conclusion had been confirmed by confirmed by experiments, which it was found that the H^+–K^+-pump has significant activity at small external potassium concentrations (Martirosov and Trchounian 1986). An alternative possibility is the robust strategy, in which the cell reduces its efficiency (increases its power consumption) to maintain an almost constant intracellular potassium concentration.

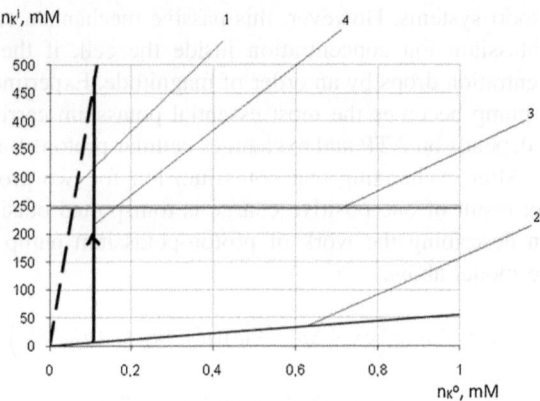

Fig. 3.7 The dependence of the intracellular potassium ion on its extracellular ion concentration with two independent transport systems. *1*—H$^+$-K$^+$-pump, *2*—passive transport, *3*—normal state, *4*—effective behavioral strategy of transport systems when concentration of environmental potassium is low (switching from one transport system to another)

As an example of the regulation of sodium ion transport, we shall consider the behavior of *Candida versatilis* (*halophila*) during salt stress (Silva-Graca et al. 2003). As noted in the article, various kinds of osmolite activity as well as enzyme activation and inactivation have been observed during salt stress. It is noted that *halophila* maintains a relative physiological equilibrium of sodium and potassium ions. The following figure shows the dependence of the intracellular sodium concentration on the environmental ion concentration.

The authors (Silva-Graca et al. 2003) did not measure the working speeds of various ion transport systems in their study. However, based on the models mentioned above, it is possible to arrive at the following interpretation of the data shown in Fig. 3.8.

There are two different sodium transport systems in *halophila*: a Na$^+$/H$^+$-antiporter and a Na$^+$-ATPase. Na$^+$/H$^+$-antiporters participate not only in sodium transport but also in the transport of lithium, potassium and rubidium (Rodriguez-Navarro et al. 1994).

The formula for the sodium ion flux for the antiporter can be written as follows:

$$J_{Na-H} = C_{Na-H}(n^i_{Na} n^o_H - n^o_{Na} n^i_H) = 0, \qquad (3.47)$$

where C_{Na-H} is a constant related to sodium ion transport out of a cell and proton transport into a cell.

Equating the sodium ion flux (3.47) to zero and using Eq. 3.36 with m = 2 for protons, we obtain:

$$n^i_{Na} = n^o_{Na} \exp\left(-\frac{\Delta \mu_A}{2} - \varphi\right). \qquad (3.48)$$

3.5 Regulation of Ion Transport in Select Microorganisms

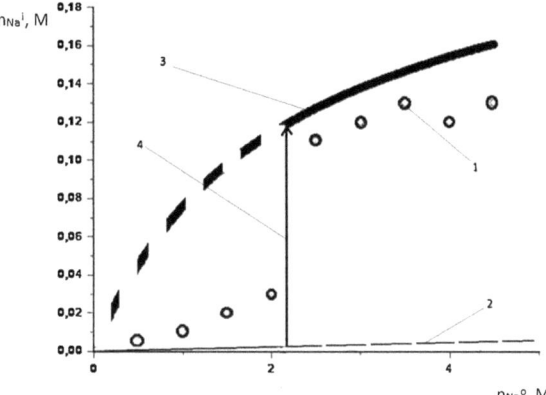

Fig. 3.8 The dependence of the intracellular sodium concentration on the environmental sodium concentration with the activity of various transport systems. *1*—experimental points (Silva-Graca et al. 2003), *2*—Na$^+$-ATPase, *3*—Na$^+$/H$^+$-antiporter, *4*—switching

The equation for the flux of sodium ions for the Na$^+$-ATPase is:

$$J_{Na} = C_{Na}\left[\exp(\Delta\mu_A + \varphi)n^i_{Na} - n^o_{Na}\right]. \quad (3.49)$$

Equating the flux to zero, we obtain:

$$n^i_{Na} = n^o_{Na}\exp(-\Delta\mu_A - \varphi). \quad (3.50)$$

Let us construct the dependences of the intracellular sodium ion concentrations on the extracellular sodium ion concentrations with sodium transport by a Na$^+$-H$^+$-exchanger and a Na$^+$-ATPase (Fig. 3.8).

As evident from the experimental data, the internal sodium concentration fundamentally changes at an external sodium concentration of approximately 2 M. The figure shows that the Na$^+$-ATPase works up to environmental sodium concentrations of 2 M, whereas, most likely, the Na$^+$/H$^+$-antiporter is switched on at environmental sodium concentration greater than 2 M.

For example, the robust strategy evidently takes place during pH regulation in yeast (Orij et al. 2011). The intracellular pH remains stable (around neutral) in response to external pH shifts in the range of 3.0 to 7.5 (Orij et al. 2009). A Na$^+$/H$^+$ antiporter, an H$^+$-ATPase and a Cl$^-$/H$^+$ antiporter were shown to be involved in pH homeostasis (Ferreira et al. 2001; Orij et al. 2011).

In the simplest case of the robust strategy, we can choose an H$^+$-ATPase as the first system and a Na$^+$/H$^+$-exchanger as the second transport system.

3.6 Possible Regulatory Strategies for Bacterial Transport of Heavy Metals

The qualitative agreement between the calculated results of these ion transport regulatory strategies and the experimental data, indicated that many microorganisms use these strategies to optimize their existence in variable environments. An analysis of bacterial transport systems that remove the heavy metal ions from a cell was performed by Silver (1996). Metal ions such as Ag^+, Cd^{2+}, Co^{2+}, Cu^{2+}, Hg^{2+}, Ni^{2+}, Pb^{2+}, Sb^{3+}, Tl^+, Zn^{2+} and others were considered in the article.

Heavy metal transport systems may be divided into two types (Silver 1996): ATPases and exchangers of bivalent ions with two protons (for example, a cadmium exchanger). The transport stoichiometry of ATPase is often variable. For example, it is known that the stoichiometry of a V-ATPase can vary from 1 to 4 protons per ATP molecule (Wagner et al. 2004). However, the transport stoichiometry of these ions is not indicated in (Silver 1996).

For the case in which an ATPase transports a single-charged ion, we have:

$$\exp(\Delta\mu_A + \varphi) = \frac{n^o_{Me}}{n^i_{Me}},$$

where n^o_{Me} and n^i_{Me} are the metal ion concentrations on the outside and the inside of a cell.

For the transport of a double-charged ion, we have:

$$\exp(\Delta\mu_A + 2\varphi) = \frac{n^o_{Me}}{n^i_{Me}}. \tag{3.51}$$

When we use the equation for protons, we obtain the following for the case in which a double-charged ion is exchanged with two protons:

$$\frac{n^o_{Me}}{n^i_{Me}} = \left(\frac{n^o_H}{n^i_H}\right)^2 = \exp(\Delta\mu_A + 2\varphi). \tag{3.52}$$

Note that Eq. 3.51 and Eq. 3.52 appeared to be identical, despite the distinction in ion transport mechanisms.

If the environmental concentrations of toxic metal ions are insignificant, the specified transport systems create a lower concentration of these metals inside of the cell than outside of the cell. However, as environmental ion concentrations increase, the work of these systems appears to be insufficient.

Note that passive transport of heavy metals must clearly be excluded from consideration as a regulatory system, as it results in the accumulation of heavy metals within the cell (because the potential inside the cell is negative).

As a hypothesis, which certainly calls for an experimental validation, let us suppose that the transport of heavy metal ions has a variable stoichiometry.

For example, if positive, single-charged ions are transported out of a cell (for example, heavy metal ions in bacteria) in a quantity of m ions per ATP molecule, we have the following equation in stationary conditions:

$$J_{Me} = C_{Me}\left(\exp(\Delta\mu_A + m\varphi)\left(n^i_{Me}\right)^m - \left(n^o_{Me}\right)^m\right) = 0.$$

3.6 Possible Regulatory Strategies for Bacterial Transport of Heavy Metals

This then leads to:

$$\exp\left(\frac{\Delta\mu_A}{m} + \varphi\right)n^i_{Me} = n^O_{Me}.$$

When the ion concentration is low, it will not influence the resting potential. It is clear, that the value of m must be reduced to lower the ion concentration inside the cell. The minimum value of m is 1. This value is generally a function of concentration. Another possibility is the use of more than one ATP molecule for the transport of each ion.

Thus, we shall assume that the cells consecutively switch from one transport system to another (with specific stoichiometries) as a strategy for counteracting high concentrations of heavy metal ions. The dependence of internal ion concentrations on external ion concentrations has shown the effectiveness of consecutively switching between transport systems.

Each cell should have a "corridor" of allowable internal ion concentrations [on Fig. 3.9, it is $(\tilde{n}^i - \delta, \tilde{n}^i + \delta)$]. The movement within this corridor can be observed as a "sawtooth" trajectory.

According to (Cornelius 1990; Goldshleger et al. 1990; Sumbilla et al. 2002), the idea of variable stoichiometries has been applied to a wide range of transporters, including V-ATPases, the Na–K-ATPase, the SR calcium pump and others.

As an alternative, a sufficiently narrow corridor of allowable internal concentrations results in the robust strategy, which only uses two transport systems (Fig. 3.10).

According to Melkikh and Sutormina (2011), both transport systems work simultaneously in the interval of concentrations between points G and F, with the internal ion concentration remaining constant and equal to $\tilde{n}^i$. However, in this case, the efficiency of the process is low enough to result in additional power consumption.

Let us note that microorganisms should have active transport systems for practically all positive ions. The necessity of active transport for almost all positive ions is due to the value of the resting potential, which is up to 8kT in

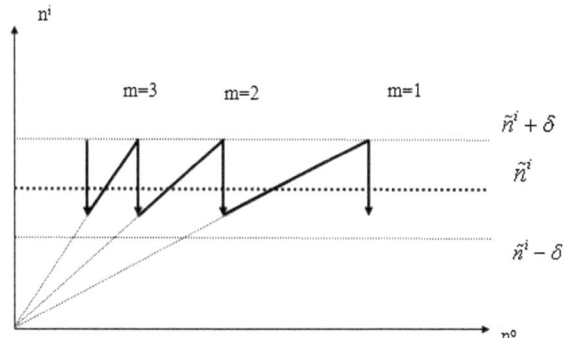

Fig. 3.9 "Sawtooth" strategy of cell behavior with strong variations in the metal concentration in an environment

Fig. 3.10 The robust strategy of cellular behavior in an environment with a highly variable metal concentration

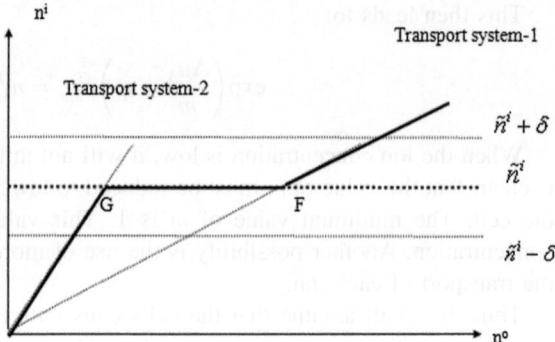

microorganisms. In this case, the concentration ratio of single-charged ions outside the cell versus inside the cell is approximately 3000. For double-charged ions, this concentration ratio is approximately 10^7. In the latter case, this means that even the concentration of microelements is too high within the cell. This, in turn, results in a falling resting potential. Probable mechanisms of active ion transport are a Na^+-ATPase or a Ca^{2+}-ATPase.

The algorithms described here can also be used for artificial cells. For example, they can be applied for the accumulation of required substances from the environment or for cellular survival in extreme conditions. However, the adaptive mechanisms of microorganisms in the extreme conditions of other planets can promote the distribution of life throughout the Universe.

3.7 Plant Cells

Plant cells have the capability of transporting ions across membranes. In particular, the leading role in the transport systems of plant cells, unlike animal cells, is played by the transport of protons. Despite the large amount of experimental data on ion transport systems in plant cell, models predicting the potential of the membrane and the concentrations of ions inside the cell are absent in the literature.

Table 3.7 presents data on the concentrations of the ions in plant cells and vacuoles (Serrano and Rodriguez-Navarro 2001).

In plant cells, the main systems of active ion transport (Sperelakis 2001; Stout and Griffin 2001) are the following:

Table 3.7 Ion concentrations in plant cells and vacuoles

Value	External media	Cytoplasm	Vacuole
pH		7.6	5.5
ΔpH		2–3	1–2
Potassium (mM)	0.1	75-100	150
Potential	0	−100–200	−80–20

3.7 Plant Cells

- H^+-ATPase, which is responsible for the transport of protons from the cell,
- Na^+-H^+ exchanger, which is an ATP-independent system for the exchange of a sodium ion for a proton, and
- Ca^{2+}-ATPase, which is responsible for the pumping of calcium out of the cell.

A scheme of the main systems of ion transport in plant cells is shown in Fig. 3.11.

Based on the algorithm described in Chap. 1, consider the main ion transport systems. The equation for the active proton flux can be written using the previously obtained formula for the flux and taking into account that n ions are transferred:

$$J_H = C_H \left[\exp(\Delta \mu_A + n\varphi)(n_H^i)^n - (n_H^o)^n \right].$$

Equating the flux to zero, we obtain the following equation for the concentration of protons:

$$n_H^i \exp\left(\frac{\Delta \mu_A}{n} + \varphi \right) = n_H^o.$$

This equation can be rewritten in the following form:

$$\frac{\Delta \mu_A}{n} = -\varphi + \ln \frac{n_H^o}{n_H^i} = \Delta \mu_H.$$

Using experimental data on the potential ($\varphi = -180$ mV) and ΔpH of the membrane (2 units) and the value of $\Delta \mu_A \approx 20$ (in dimensionless form), we obtain the following:

$$\frac{20}{n} = \frac{180}{25} + \ln 100 = 7.2 + 4.6 = 11.8.$$

It is evident that the value of n is equal to two. Thus, the number of protons can be determined:

Fig. 3.11 The scheme of the transport of ions in plant cells

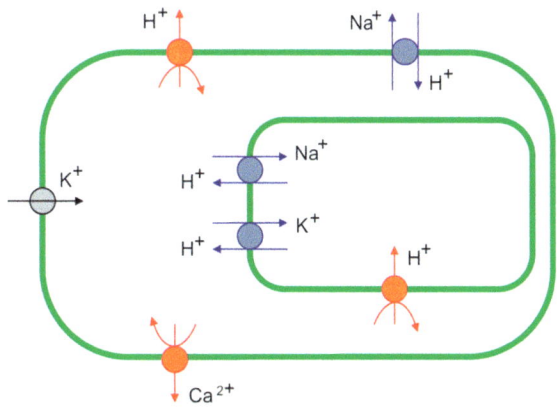

$$n_H^i \exp\left(\frac{\Delta\mu_A}{2} + \varphi\right) = n_H^o$$

The main system responsible for the transport of sodium ions in plant cells is an exchanger, which is responsible for the exchange of these ions with protons. We write the equation for the transfer of sodium ions through the Na$^+$–H$^+$–exchanger given that one sodium ion is exchanged for one proton and is thus removed from the cell:

$$J_{Na-H} = C_{Na-H}\left[n_{Na}^i n_H^o - n_{Na}^o n_H^i\right],$$

where C_{H-Na} is a constant. The dependence of the intracellular sodium ion concentration on the extracellular concentration is the following:

$$n_{Na}^i = n_{Na}^o \exp\left(-\frac{\Delta\mu_A}{2} - \varphi\right).$$

We also assume that chlorine ions are distributed passively:

$$n_{Cl}^i = n_{Cl}^o \exp(\varphi).$$

According to the literature (Sperelakis 2001), many different types of potassium channels exist. Assuming that the potassium ions are distributed passively, we can write the following formula:

$$n_K^i = n_K^o \exp(-\varphi).$$

In the equation for the calcium ions, we take into account only the work of the ATP-dependent pump. Then, the equation for calcium can be written as

$$J_{Ca} = C_{Ca}\left[\exp(\Delta\mu_A + 2\phi)n_{Ca}^i - n_{Ca}^o\right].$$

Hence, the expression for the intracellular calcium ion concentration will be

$$n_{Ca}^i = n_{Ca}^o \exp(-\Delta\mu_A - 2\varphi).$$

We can then write the condition of electroneutrality for the intracellular environment. To obtain the analytical expressions for the membrane potential, we neglect the concentration of calcium ions and other doubly charged ions. We also take into account that there are non-penetrating ions (metabolites) within the cell. Thus, we have

$$n_K^i + n_{Na}^i = n_{Cl}^i + Z_A n_A.$$

Substituting the expressions for the concentrations of ions, we obtain

$$n_K^o \exp(-\varphi) + n_{Na}^o \exp\left(-\frac{\Delta\mu_A}{2} - \varphi\right) = n_{Cl}^o \exp(\varphi) + Z_A n_A.$$

The solution of this equation can be represented in the following form:

$$\varphi = \ln\left(\sqrt{\left(\frac{Z_A n_A}{2n^o_{Cl}}\right)^2 + \frac{n^o_K}{n^o_{Cl}} + \frac{n^o_{Na}}{n^o_{Cl}}\exp\left(-\frac{\Delta\mu_A}{2}\right)} - \frac{Z_A n_A}{2n^o_{Cl}}\right). \tag{3.53}$$

Equation (3.53) can be simplified by taking into account the fact that the concentration of potassium ions in the external environment is always lower than the concentration of non-penetrating ions in the cell [the second and first terms under the square root of Eq. (3.53)]. The third term is also small at all sodium concentrations due to the effect of the exponent. Therefore, after expanding the equation in series using small quantities, we have

$$\varphi \approx \ln\left(\frac{n^o_K}{Z_A n_A}\right). \tag{3.54}$$

A similar expression was obtained earlier for *E. coli*. After substituting the numerical values of the terms, we have

$$\varphi \approx \ln\left(\frac{n^o_K}{n_A}\right) = \ln\left(\frac{0.1}{100}\right) \approx -173 mV,$$

which agrees well with the experimental data.

We now consider the dependence of the internal concentrations on the external concentrations under salt stress conditions (e.g. the addition of sodium, potassium and chlorine to the environment). In this analysis, we assume that sodium, chlorine and potassium are added in the same proportions as in the previous analysis that was performed in the absence of stress.

Under salt stress conditions in which potassium (1 mM) and sodium (100 mM) are added to the environment, the expansion of Eq. (3.53) to obtain Eq. (3.54) is valid because of the large concentration of chlorine. Using this equation, we obtain the following potential:

$$\varphi \approx \ln\left(\frac{n^o_K}{Z_A n_A}\right) = \ln\left(\frac{1}{100}\right) \approx -115 \, mV.$$

The concentration of sodium ions inside the cell is therefore

$$n^i_{Na} = n^o_{Na} \exp\left(-\frac{\Delta\mu_A}{2} - \varphi\right) = 100 \exp(-5.3) = 0.5 \, mM.$$

We also obtain the following potassium concentration:

$$n^i_K = 99.7 \, mM.$$

Thus, under salt stress conditions, the concentrations of sodium and potassium remained almost unchanged despite the significant change in the membrane potential. This property of the plant cell can be considered as a regulatory strategy. It is possible that the membrane potential in plant cells is chosen to control the concentration of ions inside the cell under salt stress conditions. In other words, the potential may simply act as a buffer that allows the cell to regulate certain ions. It is remarkable that such a strategy does not require switching between the different systems of transport.

However, this stress will change the pH ($\Delta pH = 2.3$) and reduce the concentration of calcium ions (depending on whether calcium is added in combination with sodium). Thus, the medium inside the cell becomes more acidic.

The resultant pressure inside the cell may depend on the transport of ions in the vacuole, as will be shown later.

3.8 Vacuoles

Vacuoles are membrane-bounded compartments within some eukaryotic cells that serve a variety of secretory, excretory, and storage functions. Vacuoles and their contents are considered to be distinct from the cytoplasm and are classified as ergastic by some authors (Stout and Griffin 2001; Alberts et al. 2002). Vacuoles are especially conspicuous in most plant cells.

In general, the vacuole functions include the following:

- removal of unwanted structural debris,
- isolation of materials that might be harmful to the cell,
- containment of waste products,
- maintenance of internal hydrostatic pressure (turgor) within the cell,
- maintenance of an acidic internal pH,
- containment of small molecules,
- export of unwanted substances from the cell, and
- regulation of cellular shape changes.

Vacuoles also play a major role in autophagy and maintain a balance between the biogenesis (production) and degradation (or turnover) of many substances and cell structures. They also aid in the destruction of invading bacteria and misfolded proteins that build up within the cell.

Most mature plant cells have one or several vacuoles that typically occupy more than 30 % of the volume of the cell, although vacuoles can occupy as much as 90 % of the volume in certain cell types and under certain conditions (Stout and Griffin 2001; Alberts et al. 2002). A vacuole is surrounded by a membrane called the tonoplast.

A vacuole houses large amounts of a liquid called cell sap, which is composed of water, enzymes, inorganic ions (such as K^+ and Cl^-), salts (such as calcium), and other substances, including toxic byproducts that are removed from the cytosol to avoid their interference in the cellular metabolism. The toxins located in the vacuole may also help protect some plants from predators. The transport of protons from the cytosol to the vacuole aids in the maintenance of a stable cytoplasmic pH and increases the acidity of the vacuolar interior, which in turn increases the activity of degradative enzymes. Although the presence of a large central vacuole

3.8 Vacuoles

is the most common case, the size and number of vacuoles may vary in different tissues and developmental stages. The cells of the vascular cambium, for example, have many small vacuoles in the winter and a single large vacuole in the summer.

In addition to storage, the main role of the central vacuole is to maintain the turgor pressure against the cell wall. The proteins found in the tonoplast control the flow of water into and out of the vacuole through active transport, which involves the pumping of potassium (K^+) ions into and out of the vacuolar interior. Due to osmosis, water will diffuse into the vacuole, which places pressure on the cell wall. If the water loss leads to a significant decline in the turgor pressure, the cell will plasmolyse. The turgor pressure exerted by the vacuoles is also helpful for cellular elongation: as the cell wall is partially degraded by the action of auxins, the less rigid wall is expanded by the pressure from the vacuole. In fact, vacuoles can help some plant cells reach considerable size. Another function of a central vacuole is that it pushes all of the contents of the cellular cytoplasm against the cellular membrane and thus keeps the chloroplasts closer to the light.

The vacuole also stores the flower pigments.

The potential on the vacuole membrane relative to the cytoplasm varies from approximately -80 to -20 mV (the interior of the vacuole is positively charged).

The main systems of ion transport in the vacuole include the following (Stout and Griffin 2001; Alberts et al. 2002; Grabe et al. 2000; Sze et al. 1999; Drozdowicz and Rea 2001):

- H^+-ATPase, which transports the protons into the vacuole (according to the literature, this ATPase is considered the basic transport of protons),
- Pyrophosphatase, which uses pyrophosphate to transport protons into the vacuole,
- K^+–H^+ and Na^+–H^+ exchangers (antiporters), and
- Passive channels for each type of ions.

The pH of the vacuole interior is approximately equal to 5.5 (as in the environment). Table 3.8 shows the ion concentrations in the vacuoles of various plants (De 2000).

The data shown in the table indicate that the relationship of the concentrations of the different ions between the vacuole and the cytoplasm of different plant cells is quite different (likely due to the different mechanisms of ion transport). Furthermore, chloride ions form the majority of the negative ions in the vacuole.

We will now consider the transport of the major ions in the vacuole, in accordance with (Melkikh and Seleznev 2012) and based on the previously proposed algorithms.

Protons are transported by ATPase and the exchanger systems, which exchange protons with potassium and sodium ions as well as doubly charged cations. In the case of the opposite transfer to other cations, protons act as the driving force, which means that the main mechanism for the transport of H^+ must be active. From the expression for the active flux of protons that was used previously for a variety of cells, we can obtain

Table 3.8 Concentrations of ions in the vacuoles of various plants

	Compartments	Ion concentration, mM		
		K^+	Na^+	Cl^-
Nitella translucens	Cytoplasm	119	14	65
	vacuole	75	65	150–170
Acetabularia mediterranea	Cytoplasm	400	57	480
	vacuole	355	65	480
Valonia ventricosa	Cytoplasm	434	40	138
	vacuole	625	44	643
Oat (Avena) Embryonic leaf	Cytoplasm	178	15	83
	vacuole	174	27	65

$$n_H^V = n_H^i \exp\left(\frac{\Delta\mu_A}{3} + \varphi_V\right),$$

where φ_V is the cell potential with respect to the vacuole (this is always negative) and n is the number of protons that are transferred simultaneously, which equals to 3. This number can be obtained based on the known quantities, a potential of -50 mV and a ΔpH value of 2.

The transport of potassium ions is provided by the potassium channels and the exchangers with protons. From the data in Table 3.8 and the electric potential, it can be concluded that we cannot select one transport system as the primary system of potassium transport. Indeed, if passive transport is chosen as the main system, the concentration of potassium ions inside the vacuoles would be several times lower than in the cytoplasm. The expression for the flow of potassium ions produced by an exchanger is the following:

$$J_{K-H} = C_{K-H}\left[n_K^i n_H^V - n_K^V n_H^i\right] = 0.$$

Based on this equation, we can obtain the ratio of the concentration of potassium ions using the known pH. For example, using the known values of the pH and potassium concentration in the cells of the algae *Nitella*, we find that the exchanger creates a concentration of potassium ions inside the vacuole of 11.9×10^3, which is not in agreement with reality. On the other hand, the passive flow of potassium ions in the absence of other mechanisms creates an internal concentration of approximately 16.2 mM, which also does not correspond to the experimental data.

Consequently, we can assume that these two systems work in combination to provide a constant total concentration of potassium in the vacuole.

The transport of sodium ions is the same: there are passive channels and exchangers with protons. The expressions for the fluxes generated by each exchanger separately are similar to those mentioned earlier. Accordingly, we can conclude that these two transport systems also work together to regulate the concentration of sodium ions.

3.8 Vacuoles

We now consider the effect of changing the concentration of ions in the vacuole and the cytoplasm to simulate salt stress conditions. Under salt stress conditions, the concentration of potassium in the vacuole is small, whereas the concentration of sodium is large (Serrano and Rodriguez-Navarro 2001). This can be achieved as follows: potassium ions are transported only through the passive transporter and the sodium ions are transported only through the exchanger.

However, the inositol simporter also transfers sodium from vacuoles. This transporter can also act as a regulator because, according to previous research (Serrano and Rodriguez-Navarro 2001), the inositol-exchanger and an external sodium-proton exchanger change their activity under stress conditions (regulated). In addition, the pH in the vacuole may also change in response to stress, which would also entail changes in the sodium transport.

How can vacuole regulate the pressure in a cell? At first glance, changing the volume of the vacuole cannot affect the osmotic pressure inside the cell. This conclusion stems from the fact that the pressure in the cytoplasm is completely determined by the concentrations of ions in the cytoplasm. These concentrations, in turn, are determined by the transport of the different ions through the outer cell membrane. However, this is not the case with the non-penetrating ions. As observed from Table 3.8, the non-penetrating ions are found in greater amounts in the cytoplasm than in the vacuole. The concentration of non-penetrating ions (because their number remains constant) depends on the volume that they occupy. Consequently, an increase in the vacuole volume will decrease the volume of the cytoplasm, which will increase the cytoplasmic pressure. Consequently, it is possible to regulate the cellular turgor through changes in the volume of the vacuole, which can be achieved by the regulation of certain transport systems.

3.9 Thylakoid

A model for the conversion of light energy to the free energy of the protons has been previously proposed (Melkikh et al. 2010). We describe briefly the results of the proposed model. The main feature of the conversion of light energy is that the photon has a chemical potential that is equal to zero, which means that a photon cannot be considered using the same ideology as the transported molecules. However, as shown in the previous work (Melkikh et al. 2010), the absorption of a photon by a two-level system, in which the difference in energy levels is equal to the photon energy, forces the system into a non-equilibrium state, in which the maximum difference in the chemical potential (in the absence of additional loss) equals to

$$\Delta\mu^{max} = kT\left(\frac{h\nu}{kT} - \frac{h\nu}{kT_{\nu\Omega}}\right),$$

where $T_{\nu\Omega}$—is the temperature of the Sun.

In

$$T \ll T_{\nu\Omega},$$

(which is valid for a wide range of temperatures on Earth), we receive

$$\Delta\mu^{max} \approx h\nu.$$

This two-level system, which is not in equilibrium due to the absorption of photons from the sun, is capable of transporting any substance (including protons).

In accordance with experimental data (Volkenstein 1983; Albarran-Zavala and Angulo-Brown 2007), an average efficiency in the chain of photosynthesis reactions is $\xi = 0.33$; therefore, it can be assumed that $\Delta\mu = \frac{h\nu}{3}$. In a previous study (Melkikh et al. 2010), the following equation for proton flux, which is the result of sunlight, was reported:

$$J_H = k_{H\uparrow} \frac{n_2^a n_{2o}^*}{n_2^2} \left(e^{\frac{h\nu}{3kT} - \frac{2e\varphi}{kT}} (n_H^o)^2 - (n_H^i)^2 \right),$$

where $k_{H\uparrow}$, n_2^a, n_{2o}^*—are constants.

According to experimental data, protons are carried through the thylakoid membrane, which separates the lumen and the stroma, due to photochemical reactions and the operation of ATP synthase.

Using the dimensionless variables $h\nu \to \frac{h\nu}{kT}$, $\varphi \to \frac{e\varphi}{kT}$, $\Delta\mu_A \to \frac{\Delta\mu_A}{kT}$, the equation for the proton flow can be written as

$$J_H = k_{H\uparrow} \frac{e^{\Delta\mu_2 - 2\varphi}(n_H^o)^2 - (n_H^i)^2}{(e^{\Delta\mu_2 - h\nu} + 1)(1 + e^{h\nu})}.$$

The second flow of protons, which is controlled by ATP synthase, can be defined as in a previous work (Melkikh and Seleznev 2007):

$$J_H^{ATP} = k_H^{ATP} \frac{e^{-\Delta\mu_A + m\varphi}(n_H^i)^m - (n_H^o)^m}{(1 + e^{\Delta\mu_A + m\varphi - Q})}.$$

The frequency of the ATP synthesis reaction events, ν_{ATP}, is related to the proton flow:

$$\nu^{ATP} = \frac{J_H^{ATP}}{m}.$$

At present, there is no agreement among investigators regarding the stoichiometric coefficient m of the ATP synthase-regulated process. The data varies: $2H^+$

(Hall and Rao 1981; Nicholls 1982; Rubin 1987); 3H$^+$ (Riznichenko et al. 1999); and undefined (Creighton 1999).

In mitochondria, the ATP synthase is of the same F_1F_0 type (Gennis 1989) and has a stoichiometric coefficient of $m = 3$ (Melkikh and Seleznev 2007).

A simplified transport scheme for the thylakoid membrane is shown in Fig. 3.12. The transport system includes the following elements:

- a proton pump, which feeds H^+ ions to the inside by absorbing light from the photosystem II (conformon K2),
- ATP synthase, which performs ATP synthesis based on the different electrochemical potentials of the hydrogen ions at the thylakoid membrane,
- a primary PA acceptor outside the thylakoid, which accepts an excited electron from chlorophyll on the inside through the absorption of a light quantum from the photosystem I (conformon K1), and;
- passive transport of H^+, K^+, Cl^-, A^- anions.

It should be remembered that the concentration of protons in the stroma solution is sufficiently small ($n_H^o \leq 10^{-4}$ mM) that these cannot have a considerable effect on the formation of the potential. We are unaware of any data that indicates the presence of the active transport of other anions or cations. Therefore, the only mechanism responsible for the charging of the thylakoid membrane is the transport of electrons.

In the paper (Melkikh et al. 2010), an expression for the membrane resting potential of the thylakoid membrane was presented. The resulting value of 8 mV is consistent with experimental data.

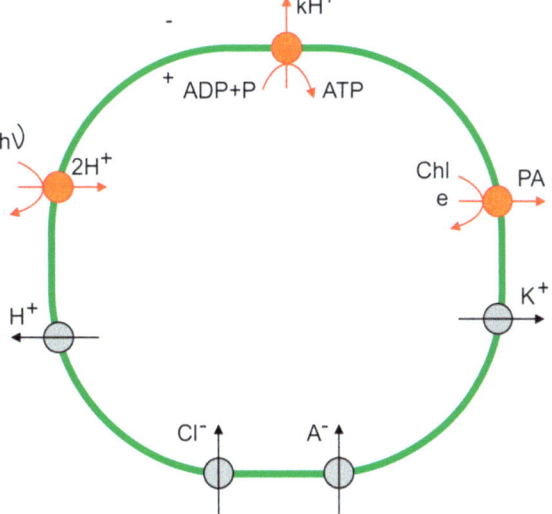

Fig. 3.12 Scheme of proton and electron flows through a thylakoid membrane. *PA*—primary acceptor

3.10 Conclusion

In this chapter, a series of modeling techniques were used to study plant cells, unicellular organisms and some cellular compartments. Based on the constructed models, it can be concluded that the maintenance of a constant internal environment and the efficiency of the transport processes is a characteristic of microorganisms and plants and animal cells. A distinctive feature of microorganisms is their ability to survive in environments with a wide range of concentrations of different substances. For example, bacteria that live in an environment containing a small amount of potassium switch on an additional ATP-dependent pump to maintain their internal potassium concentration. Another feature of many microorganisms is the ability of their cell walls to withstand the large pressure difference between the cytoplasm and the external environment. In some cases (e.g. plants), the pressure inside the cell can be maintained through the regulation of the vacuolar transport systems.

References

Albarran-Zavala E, Angulo-Brown F (2007) A simple thermodynamic analysis of photosynthesis. Entropy 9:152–168
Alberts B, Johnson A, Lewis J, Raff M, Roberts K, Walter P (2002) Molecular biology of the cell, 4th edn. Garland Science, New York
Bakker EP, Rottenberg H, Caplan SR (1976) An estimation of the light-induced electrochemical potential difference of protons across the membrane of Halobacterium halobium. Biochim Biophys Acta 440(3):557–572
Bara M, Guiet-Bara A, Durlach J (1993) Regulation of sodium and potassium pathways by magnesium in cell membranes. Magnes Res 6(2):167–177
Bhattacharyya P, Volcani BE (1980) Sodium-dependent silicate transport in the apochlorotic marine diatom Nitzschia alba. Proc Natl Acad Sci 77(11):6386–6390
Boalch GT (1987) Changes in the phytoplankton of the western English Channel in recent years. Br Phycol J 22:225–235
Bogomolni RA (1977) Light energy conservation processes in Halobacterium halobium cells. Fed Proc 36(6):1833–1839
Borrelly G, Boyer JC, Touraine B, Szponarski W, Rambier M, Gibrat R (2001) The yeast mutant vps5Delta affected in the recycling of Golgi membrane proteins displays an enhanced vacuolar Mg^{2+}/H^+ exchange activity. Proc Natl Acad Sci 98(17):9660–9665
Boyd CM, Gradmann D (1999a) Electrophysiology of the marine diatom Coscinodiscus wailesii I. Endogenous changes of membrane voltage and resistance. J Exp Bot 50:445–452
Boyd CM, Gradmann D (1999b) Electrophysiology of the marine diatom Coscinodiscus wailesii II. Potassium currents. J Exp Bot 50:453–459
Boyd CM, Gradmann D (1999c) Electrophysiology of the marine diatom Coscinodiscus wailesii III. Uptake of nitrate and ammonium. J Exp Bot 50:461–467
Briskin DP (1990) The plasma membrane H^+-ATPase of higher plant cells: biochemistry and transport function. Biochem Biophys Acta 1019(2):95–109
Brownlee C, Wood JW, Briton D (1987) Cytoplasmic free calcium in single cells of centric diatoms. Protoplasma 140(2–3):118–122

References

Cornelius F (1990) Variable stoichiometry in reconstituted shark Na, K-ATPase engaged in uncoupled efflux. Biochim Biophys Acta 1026:147–152
Creighton TE (1999) Encyclopedia of molecular biology. Wiley, New York
De DN (2000) Plant cell vacuoles: an introduction. Csiro Publishing, Collingwood
Detkova EN, Pusheva MA (2006) Energy metabolism in halophilic and alkaliphilic bacteria. Microbiology 75(1):5–17
Drozdowicz YM, Rea PA (2001) Vacuolar H^+ pyrophosphatases: from the evolutionary backwaters into the mainstream. Trends Plant Sci 6(5):206–211
Eisenbach M (1982) Changes in membrane potential of Escherichia coli in response to temporal gradients of chemicals. Biochemistry 21(26):6818–6825
Ferreira T, Mason AB, Slayman CW (2001) The yeast Pma1 proton pump: a model for understanding the biogenesis of plasma membrane proteins. J Biol Chem 276:29613–29616
Garrett RH, Grisham CM (2002) Biochemistry, 2nd edn. Brooks/Cole, Pacific Groove
Gennis RB (1989) Biomembranes: molecular structure and function. Springer, New York
Goldshleger R, Shahak Y, Karlish SJD (1990) Electrogenic and electroneutral transport modes of renal Na/K ATPase reconstituted into proteoliposomes. J Membr Biol 113:139–154
Grabe M, Wang H, Oster G (2000) The mechanochemistry of V-ATPase proton pumps. Biophys J 78:2798–2813
Gradmann D, Blatt MR, Thiel G (1993) Electro coupling of ion transporters in plants. J Membr Biol 136(3):327–332
Gradmann D, Boyd CM (2000) Three types of membrane excitations in the marine diatom *Coscinodiscus wailesii*. J Membr Biol 175(2):149–160
Grogan DW (2001) Physiology of procaryotic cells. In: Sperelakis N (ed) Cell physiology sourcebook, 3rd edn. Academic, San Diego
Hall DO, Rao KK (1981) Photosynthesis. Edward Arnold and Co, London
Ingraham JL, Maaloe O, Neidhardt FC (1983) Growth of the bacterial cell. Sinauer Association, Sunderland
Kinclova O, Potier S, Sychrova H (2002) Difference in substrate specificity divides the yeast alkali-metal-cation/HM antiporters into two subfamilies. Microbiology 148:1225–1232
Lanyi JK (1978) Light energy conversion in *Halobacterium halobium*. Microbiol Rev 42(4):682–706
Martirosov SM, Trchounian AA (1986) 838: An electrochemical study of energy-dependent potassium accumulation in E. coli: Part XI. The Trk system in anaerobically and aerobically grown cells. Bioelectroch Bioener 15(3): 417–426
Melkikh AV, Bessarab DS (2010) Model of active transport of ions through diatom cell biomembrane. Bull Math Biol 72(7):1912–1924
Melkikh AV, Seleznev VD (2007) Nonequilibrium statistical model of active transport of ions and ATP production in mitochondria. J Biol Phys 33(2):161–170
Melkikh AV, Seleznev VD (2009) Model of active transport of ions in archaea cells. Bull Math Biol 71(2):383–398
Melkikh AV, Seleznev VD (2012) Mechanisms and models of the active transport of ions and the transformation of energy in intracellular compartments. Prog Biophys Mol Bio 109(1–2):33–57
Melkikh AV, Sutormina MI (2011) Algorithms for optimization of the transport system in living and artificial cells. Syst Synth Biol 5(1–2):87–96
Melkikh AV, Seleznev VD, Chesnokova OI (2010) Analytical model of ion transport and conversion of light energy in chloroplasts. J Theor Biol 264:702–710
Michel H, Oesterhelt D (1976) Light-induced changes of the pH gradient and the membrane potential in *H. halobium*. FEBS Lett 65(2):175–178
Nanninga N (1985) Molecular cytology of *Escherichia coli*. Academic, London
Nass R, Cunningham KW (1997) Intracellular sequestration of Sodium by a novel Na^+/H^+ exchanger in yeast is enhanced by mutations in the plasma membrane H^+-ATPase. J Biol Chem 272:26145–26152
Neidhardt FC (1987) *Escherichia coli* and *Salmonella typhimurium*: cellular and molecular biology, vol 1. American Society for Microbiology, Washington

Nicholls DG (1982) Bioenergetics: an introduction to the chemiosmotic theory. Academic, London

Ono A, Tada K, Ichimi K (2006) Chemical composition of *Coscinodiscus wailesii* and the implication for nutrient ratios in a coastal water, Seto Inland Sea, Japan. Mar pollut bull 57(1–5):94–102

Oren A (1999) Bioenergetic aspects of halophilism. Microbiol Mol Biol Rev 63(2):334–348

Orij R, Postmus J, Ter Beek A, Brul S, Smits GJ (2009) In vivo measurement of cytosolic and mitochondrial pH using a pH-sensitive GFP derivative in *Saccharomyces cerevisiae* reveals a relation between intracellular pH and growth. Microbiol 155:268–278

Orij R, Brul S, Smits GJ (2011) Intracellular pH is a tightly controlled signal in yeast. Biochim Biophis Acta 1810:933–944

Padan E, Zilberstein D, Rottenberg H (1976) The proton electrochemical gradient in *Escherichia coli* cells. Eur J Biochem 63(2):533–541

Padan E, Bibi E, Ito M, Krulwich TA (2005) Alkaline pH homeostasis in bacteria: new insights. Biochim Biophis Acta 1717:67–88

Riznichenko G, Lebedeva G, Demin O, Rubin A (1999) Kinetic mechanisms of biological regulation in photosynthetic organisms. J Biol Phys 25:177–192

Rodrigues-Navarro A, Quintero FJ, Garciadeblás B (1994) Na-ATPases and Na^+/H^+ antiporters in fungi. Biochim Biophis Acta 1187(2):203–205

Rubin AB (1987) Biophysics. Visshaja shkola, Moscow (in Russian)

Schafer G, Engelhard M, Muller V (1999) Bioenergetics of the Archaea. Microbiol Mol Biol Rev 63(3):570–620

Serrano R, Rodriguez-Navarro A (2001) Ion homeostasis during salt stress in plants. Curr Opin Cell Biol 13(4):399–404

Silva-Graca M, Neves L, Lucas C (2003) Outlines for the definition of halotolerance/halophily in yeasts: Candida versatilis (halophila) CBS4019 as the archetype? FEMS Yeast Res 3(4):347–362

Silver S (1996) Bacterial resistances to toxic metal ions—a review. Gene 179(1):9–19

Smirnov AV, Suzina NE, Kulakovskaia TV, Kulaev IS (2002) Magnesium orthophosphate, a new form of reserve phosphate in the halophilic archaeon *Halobacterium salinarium*. Microbiologiia 71(6):786–793

Sperelakis N (2001) Cell physiology sourcebook, 3rd edn. Academic, San Diego

Stout RG, Griffing LR (2001) Plant cell physiology. In: Sperelakis N (ed) Cell physiology sourcebook, 3rd edn. Academic, San Diego

Sumbilla C, Lewis D, Hammerschmidt T, Inesi G (2002) The slippage of the Ca^{2+} pump and its control by anions and curcumin in skeletal and cardiac sarcoplasmic reticulum. J Biol Chem 277(16):13900–13906

Syrchova H (2004) Yeast as a model organism to study transport and homeostasis of alkali metal cations. Physiol Res 53(1):91–98

Sze H, Li X, Palmgen MG (1999) Energization of plant cell membranes by H^+-pumping ATPases: regulation and biosynthesis. Plant Cell 11:677–689

Ventosa A, Nieto JJ, Oren A (1998) Biology of moderately halophilic aerobic bacteria. Microbiol Mol Biol Rev 62(2):504–544

Volkenstein MV (1983) General biophysics. Academic, New York

Wagner CA, Finberg KE, Brenton S, Marshansky V, Brown D, Geibel JP (2004) Renal vacuolar H^+-ATPase. Physiol Rev 84(4):1263–1314

Chapter 4
Optimization of the Transport of Substances in Cells

We then consider the problem of optimizing the transport subsystem of artificial cells and propose an algorithm for the synthesis of an artificial transport system that can provide the specified concentration within the artificial cell and the desired resting potential. In addition a method that allows the between the internal and external concentrations is proposed. In the limit of a large number of transport systems, which do not differ significantly from each other in their properties, it is possible to simultaneously achieve constancy in the internal environment and an efficiency, arbitrarily close to 100 %. A generalized cell model in distilled water that maintains was built. The mechanisms by which chemical energy is converted into directed motion of protocells in the early stages of evolution were considered. Possible strategies of directed motion of protocells were received. These strategies include the direct conversion of light energy into mechanical energy. We propose a special stage in the evolution of protocells: a minimal moving cell. We also consider the unsolved problem of the transport of large molecules in living and artificial cells.

4.1 Optimization Methods Used for Models of Transport Subsystems of Living and Artificial Cells

The use of optimization methods are an indispensable part of almost every branch of mathematics. Many problems are reduced to finding the optimum (minimum or maximum) value of a function (functional). The development of these methods is largely predetermined by the needs of technology. Without exaggeration, we can claim that all of the technical devices created by man are optimal in some way. Throughout history, there is a continuous process of creating new technical devices, which either perform the same job as an existing device more cheaply, or perform more work than an existing device at the same cost.

Nature works in a similar way. In the evolution of organisms, a number of properties arose that gave the organism an advantage in its struggle for existence. Many systems of the organism (cells) are optimized to varying degrees. This raises questions such as: "Can we find a better structure for a system of the organism?", "If the organisms are not optimally arranged, then why?", "Are there any general optimization laws in living systems, that are guided by the same principles that would make it possible for such a system to be created artificially?". In the context of this book, we will focus on the third question. We are interested in the most general laws that govern the optimization of the transport subsystems of cells. These laws will help us understand the evolution of this system and will also help us create artificial cells that best fit the requirements. Appendixes 1 and 2 describe the main optimization methods that are used in this book.

4.1.1 Effectiveness of the Energy Conversion in the Transport of Substances Through Biomembranes

One of the most important questions in the modeling of artificial and living cells is their effectiveness, in particular the effectiveness of the transport subsystem. This is because the effectiveness (or any other criterion that evaluates the quality of the system) is an essential feature of any engineering problem, and is used in the optimization of any system. Therefore, we consider the most common hypothesis for the efficiency of the energy conversion in the transport of substances. The formula (1.12)

$$\eta = \frac{J' \cdot \Delta\mu}{v' \cdot \Delta\mu_A}.$$

for the efficiency of ion transport can be written in a more general form (1.14):

$$\eta_i = \frac{\sum_{i=1}^{n} C_i(\exp(\Delta\mu_{Ai}) - 1) \ln \frac{\sum_{i=1}^{n} C_i \exp(\Delta\mu_{Ai})}{\sum_{i=1}^{n} C_i}}{\sum_{i=1}^{n} C_i(\exp(\Delta\mu_{Ai}) - 1)\Delta\mu_{Ai}}.$$

This formula represents the ratio of useful power that could be obtained in a system with equal concentrations of ions on both sides of the membrane to the consumed power. The consumed power was also calculated assuming equal concentrations of the ions on both sides of the membrane. This formula was chosen because the steady-state flux of the ions is zero in the absence of passive transport (for example, for one transport system). At the same time, ATP would not be spent. Therefore, the calculation of the net power in this mode does not make sense.

4.1 Optimization Methods Used for Models of Transport

The active transport of ions is usually required to *maintain* a quantity (concentration or potential). In fact, if a substance is transported in a nominal mode, then its flux will not be zero. These two cases require separate modeling.

In the first case, the concentration of the transported substances (ions) is *maintained* at a certain level by active transport. We note that a specific ion concentration or electrical potential cannot perform any work. The work can be accomplished only through the difference in the concentrations or potentials. We show that, to maintain a certain concentration of ions inside the cell (or compartment), it is advantageous to have an effective system of transport, i.e., a transport system with minimal losses. Suppose that the active transport of ions occurs in addition to the passive flux. The total flux is then

$$J = C_A\left(n^i \exp(\Delta\mu_A) - n^o\right) + P\left(n^i - n^o\right) = 0, \tag{4.1}$$

where P—is the passive permeability of the membrane.

It can then be shown that the difference in the ion concentrations that is created will be lower than in the absence of passive transport:

$$n^i = n^o \frac{C_A + P}{C_A \exp(\Delta\mu_A) + P} > n^o \exp(-\Delta\mu_A), \tag{4.2}$$

because $1 > \exp(-\Delta\mu_A)$.

In this case, if a substance is pumped out of cells, then the concentration within the cell is less than the concentration outside. Therefore, the increase in the concentration on one side of the membrane due to passive transport would reduce the total concentration difference.

However, energy will be consumed continuously in the presence of passive transport, whereas energy will not be spent in its absence. Thus, the presence of passive transport leads to a permanent channel of free energy leakage. In this case, the value of $\eta < 1$ characterizes the presence of this loss.

Furthermore, Eq. (1.14) can also characterize a non-stationary situation. If the initial concentrations on both sides of the membrane are equal, the steady state is reached fastest in the absence of loss (i.e., $\eta = 1$). All other things being equal, the minimization of the response time to a changed environment can be a winning factor. In addition, due to the concentration difference created in the non-stationary case, the work can be accomplished if the energy source (e.g., ATP) is either absent or substantially reduced.

Now consider the case when the concentration of a substance (ion) is supported by two or more systems of active transport. In this case, it is important to clarify the purpose of active transport. Is it just to maintain the concentration of the transported ion or is energy recovery also a target?

At steady state, the overall flow of an ion, which is equal to the sum of all of the flows that are generated by each transport system, is equal to zero. This means that each individual flux is not equal to zero (we consider this situation an example of proton transfer):

$$J = C_1\left[n_H^i - n_H^o e^{\Delta\mu_1}\right] + C_2\left[n_H^i - n_H^o e^{\Delta\mu_2}\right] = 0. \tag{4.3}$$

The exception is the degenerate case in which the driving forces (the differences between the chemical potentials of substances by which the ions are transported) are equal $\Delta\mu_1 = \Delta\mu_2$. Then, we have

$$J_1 = C_1\left[n_H^i - n_H^o e^{\Delta\mu_1}\right] = 0, \ J_2 = C_2\left[n_H^i - n_H^o e^{\Delta\mu_2}\right] = 0. \tag{4.4}$$

Each of the two chemical reactions by which the transport occurs is coupled with ion transport. Therefore, if the total flow is equal to zero, and there are non-zero flows (J_1 and J_2) the flow of each of the chemical reactions is also not equal to zero. The concentration of ions that is created by two transport systems is the following:

$$n_H^i = n_H^o \frac{C_1 e^{\Delta\mu_1} + C_2 e^{\Delta\mu_2}}{C_1 + C_2}. \tag{4.5}$$

Each of fluxes can be written as the following:

$$J_1 = C_1 C_2 n_H^o \left[\frac{e^{\Delta\mu_2} - e^{\Delta\mu_1}}{C_1 + C_2}\right], \ J_2 = C_1 C_2 n_H^o \left[\frac{e^{\Delta\mu_1} - e^{\Delta\mu_2}}{C_1 + C_2}\right].$$

If the first force is greater than the second, then the first flux will be negative and the second is positive; the reverse is also true. In this case, each flux can perform the work. The key question then becomes whether such work is needed. If such work is not required, then it cannot be considered useful work. Then, the efficiency, which is of the form

$$\eta = \frac{C_1\left(1 - e^{\Delta\mu_1}\right) + C_2\left(1 - e^{\Delta\mu_2}\right)}{C_1(1 - e^{\Delta\mu_1})\Delta\mu_1 + C_2(1 - e^{\Delta\mu_2})\Delta\mu_2}\Delta\mu,$$

will have the sense of a power ratio, which can be obtained as the ratio of the power of ion transport when the ion concentrations are equal to the power of the separate work of pumps under the same conditions.

Thus, each of the compounds synthesized (these have a nonzero flux), must either be used in another process, synthesized in another process, or passively diffused into (from) the environment. In the latter case, as is the case with ions, useful work cannot be accomplished. In the case of an outside process, the efficiency will take a different form: the input power will be the power of one substance and the output power will be the power of the other substance. Such efficiency is not imaginary (virtual) and makes sense in the mode that is being considered. If there are any additional processes, the flow of each substance in the stationary state is again equal to zero. These additional processes must be taken into account explicitly in the expressions for the fluxes of the substances. For example, if the first force is maintained, then the second adjusts to it. In this case, the expression for the second force can be obtained from the law of conservation of the second substance:

4.1 Optimization Methods Used for Models of Transport

$$J_2 = C_1 C_2 n_H^o \left(\frac{e^{\Delta \mu_1} - e^{\Delta \mu_2}}{C_1 + C_2} \right) + J_2^o = 0,$$

where the second term is the flux of this second substance, which is caused either by its reverse reaction (passive) or an external useful process. From this equation, we can find the value of $\Delta \mu_2$.

The general rule is that all substances that are transported are subject to some form of the conservation laws. The expression for this efficiency is the following:

$$\eta = \frac{J_2 \Delta \mu_2}{J_1 \Delta \mu_1}.$$

In this situation, the protons (or any other ions) act as a link (similar to gears in a mechanical transmission) between the input and the output in the system. Because the flows J_1 and J_2 are equal we receive

$$\eta = \frac{\Delta \mu_2}{\Delta \mu_1}.$$

If the passive flow of the second substance is absent (this is equivalent to the situation in which an active flow is equal to zero or that it is useful or fully consumed), we have

$$J_2 = C_1 C_2 n_H^o \left(\frac{e^{\Delta \mu_1} - e^{\Delta \mu_2}}{C_1 + C_2} \right) = 0$$

and $\eta = 1$.

If there are several transferred ions and only one difference in chemical potential, which acts as the driving force of the transport, the analysis can follow the analysis of the case with only one ion, i.e., when the concentration of an ion is supported but there is no work performed. In this sense, the ions are only potentially able to perform the work.

Note that all of the work that is performed in living cells is to ensure that the greatest number (ceteris paribus) of copies of the organism (cell) is produced. According to the classical results of population dynamics, the dominant species (quasispecies) of an organism will win, which means that the organism will be capable of producing more copies during its lifetime. Consequently, the most effective system of transport would be one that can perform at a maximum output of useful work per unit time, which is simply consumed to make copies of the cell.

In this formulation of the problem (the same input conditions), we cannot introduce efficiencies or limit the maximum net power.

This leads to the formulation of another question: how does the output power depend on the concentration? The fact that the majority of biochemical reactions are catalytic (i.e., their rates are highly dependent on the concentration of any substance) is well known. This dependence can be determined from experimental data. If we bear in mind that the maximum power output is found with some ion concentrations, this contribution can be approximated as a quadratic term. Consequently, the minimized value for the transport of substances can be represented as

$$W = J_A \Delta \mu_A - \sum_i J_i \Delta \mu_i + \sum_k \alpha_k \left(n_k^i - \tilde{n}_k^i \right) \to \min. \qquad (4.6)$$

The first term represents the power expended, whereas the second term corresponds to the amount of useful work. The third term represents the losses that arise due to any deviation from the optimal concentration of the internal values.

In many cases, the losses caused by the difference between the internal concentration values from the desired concentrations can be consider in the relative form

$$\sum_k \alpha_k \left(\frac{n_k^i - \tilde{n}_k^i}{\tilde{n}_k^i} \right)$$

because many of the ions with a low concentration (such as protons or calcium ions) play a major role in the regulation of cellular processes.

In the analysis of the cells of an organism (or artificial cells) that only supports the internal concentration of the transported substances, the optimization function can be written without the second term:

$$W = J_A \Delta \mu_A + \sum_k \alpha_k \left(n_k^i - \tilde{n}_k^i \right) \to \min.$$

What are the losses that are associated with the deviation of the internal concentrations from the optimal values? In each case, the mechanism of the loss requires separate modeling. In general, we can state that this type of deviation involves an inefficient degree of freedom in the transformation of energy, i.e., part of the free energy is lost to the environment.

The formula for the efficiency can be generalized to a continuous distribution of the properties of transport systems. We characterized the difference between different individual transport systems with the quantity x, which can be regarded as practically continuous. Then, by introducing the distribution function of the intensity of the work of transport systems, $f(x)$, we can write the following formulas for the efficiency:

$$\eta = \frac{\ln \int f(x)(\exp(\Delta\mu(x)))dx \int f(x)(\exp(\Delta\mu(x)) - 1)dx}{\int f(x)(\exp(\Delta\mu(x)) - 1)\Delta\mu(x)dx}.$$

or

$$\eta = \frac{\ln \int f(x)g(x)dx \int f(x)(g(x) - 1)dx}{\int f(x)(g(x) - 1)\ln g(x)dx},$$

where

$$g(x) = \exp(\Delta\mu(x)).$$

4.1 Optimization Methods Used for Models of Transport

It is easy to understand that, if there is only one transport system, then

$$f(x) = \delta(x - x_0)$$

and

$$\eta = 1.$$

In the other limit case, in which all of the transport systems are equal to each other,

$$g(x) = g.$$

This also results in the efficiency being equal to one.

4.1.2 Synthesis of the Transport System of an Artificial Cell Based on the Method of Dynamic Programming

We can begin modeling the transport system of artificial cells with a constant medium, which has constant properties (concentration of ions in it). At first glance, the modeling of the cell in such a medium is a simple task that can be solved on the basis of the algorithm "one ion—one transportation system". That is, to find the transport system for each ion that is able to create the necessary concentration within the cell. The problem, however, is that the stoichiometry of ion transport can only be an integer and it is likely to be small. What if none of the integers give the desired concentration? It turns out that this problem is largely similar to the well-known problems packing and the problem of optimal coding in the absence of interference (see below). The essence of the problem of packing (e.g. a knapsack) is that we want to load a container, which that has a limited carrying capacity, with objects that have different values such that the total value in the container is maximal (see Appendix 1 for details). One of the simplest algorithms for solving this problem is a greedy algorithm, in which the container is filled with items that are consistent with the most specific value.

Considering the synthesis of the transport subsystem, which is a similar problem, we need to enumerate the transport systems and the number of transported ions to reach the required concentrations and potential values. The analogue in this case for the limited weight of the container may be the limited capacity of the genome that encodes the various transport systems.

Then, the packing problem for the transport systems can be formulated as follows: to determine the ion transport systems in a cell with a limited genomic capacity (the total number of transport systems) that would ensure maximum integral efficiency. In practice, this means that, for a given total number of different channels or pumps, we need to determine the composition. In accordance with the greedy algorithm, we will consistently first add the transport system that gives the largest contribution to the objective function.

The algorithmic sequence of steps for the synthesis of the transport system is the following:

1. Specify the desired concentrations of ions inside the cell.
2. Determine the logarithm of the ratio of concentrations for all specified ions.
3. Specify a list of the potential systems of transport.
4. Determine the stoichiometry of the transport of each ion in each transport system.
5. Find the deviation of the internal ion concentrations from the desired concentration and determine the losses associated with the deviation.
6. Reduce the losses considering the generalized pump and find its stoichiometric coefficients.
7. Begin with the synthesis of the ions for a given number of transport systems that makes the largest contribution to the objective function.

Consider the transfer of two uncharged substances A and B. Let the concentration of these compounds have the following numbers:

Desired concentrations inside the cell, mM	Concentrations in the environment, mM
$\tilde{n}_A^i = 10$	$n_A^o = 100$
$\tilde{n}_B^i = 1$	$n_B^o = 100$

The right column shows the concentrations of the substances A and B that exist in the environment, whereas the left shows the optimal concentrations that improve the efficiency of the energy conversion processes in the cell. In addition, let $\Delta\mu_A = 20$.

In accordance with the algorithm, we first define the logarithm of the ratio of concentrations:

$$\ln\frac{n_A^o}{\tilde{n}_A^i} = 2.3, \quad \ln\frac{n_B^o}{\tilde{n}_B^i} = 4.6.$$

We then find the stoichiometry of the transport of each ion based on its transport system:

$$m_A = \frac{20}{2.3} = 8.70, \quad m_B = \frac{20}{4.6} = 4.35.$$

Rounding to the nearest integer, we obtain $m_A = 9$ and $m_B = 4$. As a result, we calculate the concentrations of both substances corresponding to the integer stoichiometry: $n_A^i = 13.5$ and $n_B^i = 0.67$.

We can then determine the relative losses:

$$W_A = \alpha_A \left(\frac{\tilde{n}_A^i - n_A^i}{\tilde{n}_A^i}\right)^2 \text{ and } W_B = \alpha_B \left(\frac{\tilde{n}_B^i - n_B^i}{\tilde{n}_B^i}\right)^2.$$

4.1 Optimization Methods Used for Models of Transport

The coefficients α_A and α_B are determined from experimental data. However, if, for example, we consider these to be equal to unity, then we obtain the following total loss:

$$W = \sqrt{\left(\frac{\tilde{n}_A^i - n_A^i}{\tilde{n}_A^i}\right)^2 + \left(\frac{\tilde{n}_B^i - n_B^i}{\tilde{n}_B^i}\right)^2} = 0.48.$$

If these losses exceed the maximum allowable, then we introduce a new transport system in an attempt to reduce these losses. We therefore consider a generalized pump:

$$\left(n_A^i\right)^m \left(n_B^i\right)^k \exp(\Delta\mu_A) = \left(n_A^o\right)^m \left(n_B^o\right)^k.$$

The coefficients m and k can take negative values (the ion is transferred into the cell rather than out of it). The value of $\Delta\mu_A$ may be zero, which indicates the use of an exchanger and not a pump. Taking the logarithm of this equation and substituting the desired concentration, we obtain the following equation:

$$m\ln\left(\frac{n_A^o}{n_A^i}\right) + k\ln\left(\frac{n_B^o}{n_B^i}\right) = \Delta\mu_A.$$

We are interested in the integer solutions of this equation. Note that, for large m and k and a large number of ATP molecules, we can always find a whole solution that gives results that are arbitrarily close to the desired output. If, however, we are limited to small numbers (this corresponds to a small number of genes used for transport systems), we obtain:

$\Delta\mu_A = 20$	
k	m
4	1
3	2
3	3
2	4

$\Delta\mu_A = 0$	
k	m
−0.5	1
−1	2
−3/2	3
−2	4

Thus, through enumeration (this is realistic for small numbers), the problem can be solved.

Note that if an exchanger is used for one of the ions as the sole transport system of that ion, then the second ion can choose another exchanger. Because the flux of the first exchanger is equal to zero in the steady state, this flux will not contribute to the flux of the second ion. This issue will be discussed in Sect. 4.1.5.

What fundamentally changes in the algorithm if the presence of charged ions is allowed? First, the ions cannot, in principle, be independent of each other because of the electric potential. Second, the desired internal concentrations must also establish neutrality. Consider the synthesis of the transport systems of charged ions using, for example, a combination of Na$^+$, K$^+$ and Cl$^-$. Suppose that the medium has the following composition:

Desired concentration inside the cell, mM	Concentration outside the cell, mM
$\tilde{n}^i_{Na} = 10$	$n^o_{Na} = 100$
$\tilde{n}^i_K = 100$	$n^o_K = 10$
	$n^o_{Cl} = 110$

We can then determine the logarithm of the ratio of concentrations without the potential:

$$\ln \frac{n^o_{Na}}{\tilde{n}^i_{Na}} = 2.3 \quad \text{and} \quad \ln \frac{n^o_K}{\tilde{n}^i_K} = -2.3.$$

We can first assume that the potassium is transferred passively. We then find the stoichiometry of the transport of each ion in the transport system:

$$m_{Na} = \frac{20}{2.3} = 8.70 \quad \text{and} \quad m_K = \frac{20}{2.3} = 8.7.$$

Rounding to the nearest integer, we obtain $m_{Na} = 9$ and $m_K = 9$. As a result, we can calculate the concentration of sodium ions based on the integer stoichiometry: $n^i_{Na} = 13.5$. We can assume that the chloride ions are also transferred passively. In addition, we can add the concentration of non-penetrating ions in the cell such that $n_A = 50$.

We can then calculate the potential from the electroneutrality condition:

$$n^i_K + n^i_{Na} = n^i_{Cl} + n_A,$$

$$n^o_K \exp(-\varphi) + n^o_{Na} \exp(-\Delta \mu_A - m_{Na}\varphi) = n^o_{Cl} \exp(\varphi) + n_A.$$

From this equation, we can numerically find the electric potential as a function of the external concentrations.

In the next step, we can determine the ion concentrations inside the cell taking into account the calculated electric potential. If these concentrations do not differ much from the desired concentrations, the synthesis procedure ends; however, if the difference is large, then we continue to the next step of the algorithm in which we consider a generalized pump with charged ions:

$$\left(n^i_{Na}\right)^m \left(n^i_K\right)^k \exp(\Delta \mu_A + (m+k)\varphi) = \left(n^o_{Na}\right)^m \left(n^o_K\right)^k.$$

4.1 Optimization Methods Used for Models of Transport

As a result, we find, through the enumeration of integer values for m and k, which set of values gives the internal concentrations that are closest to the desired concentration and provides the lowest total number of transport systems (the minimum number of genes). Repeating this procedure iteratively, we can find the optimal transport system that gives internal ion concentrations that are arbitrarily close to the desired concentrations.

Note that there must be some limit on the total number of simultaneously transferred ions of one sign due to their Coulomb repulsion. As shown in Chaps. 2 and 3, there are no currently known exchangers that can transport more than five ions of different signs.

The proposed procedure can easily be supplemented by the requirement to maintain a certain pressure difference between both sides of the membrane.

It should be noted that the proposed algorithm can be attributed to the method of dynamic programming. Dynamic programming is a way to solve complex problems by breaking them into simpler subtasks. The key concept behind dynamic programming is the following: to solve a problem, it is necessary to solve some of the tasks (subtasks) and then combine the solutions of the sub-problems to obtain one common solution. Often, many of these subtasks are the same. The approach of dynamic programming is to solve each sub-problem only once, thereby reducing the amount of total computation. This is especially useful in cases in which the number of repeated sub-problems is exponentially large. More information on the method of dynamic programming is provided in Appendix 1. In the case of transport systems this method means we only need to solve the problem of the generalized pump once and can then use the solution throughout the algorithm. For example, the packing problem (Appendix 1) can be solved exactly using dynamic programming.

Note that the problem of synthesizing an artificial cell transport system is in many ways analogous to the problem of optimal coding in information theory. The analogues for the primary (coded) alphabet in this case are the internal and external ion concentrations and the analogues for the secondary alphabet are the transport systems. From coding theory, it is known that long codes can be used efficiently (with little or no redundancy of rounding) to encode any message in the absence of losses. In our problem, we see a similar situation: the use of a large number of transporters that carry some of the ions can achieve the required concentration and potential.

The analogy with coding can be extended to the case of errors. There are error-correcting codes, the essence of which is that each message contains "extra" bits (does not carry useful information) that, through the use of a particular algorithm (e.g., using the Hamming code), can be utilized to correct any errors. A similar role is played by the regulatory system of transport (see Chaps. 2 and 3), the precise role of which is the correction of the internal ion concentrations obtained through the major transport systems.

How to take into account any possible changes in the environmental concentrations for the synthesis of artificial cell transport systems? It is crucial to know a priori how these changes can affect the external concentration.

The model of two environments.
The simplest situation is the following: imagine that the external environment of the cell can only be in one of two states with known properties (e.g., fresh and salt water). Then, the probability of finding the cell in the first medium is equal to p, whereas the probability of finding the cell in the second medium is $(1-p)$.

Consider a possible sequence of syntheses of a cell transport system. First, in accordance with the algorithm discussed previously, we can find the optimal transport system for each environment separately. Then, the problem can be viewed as a game in which the two media can be considered the strategy of one player (nature) and the operating transport systems can be considered the strategies of the cell. One of the simplest variants of the mixed strategies in this case is the presence of regulators, which switch the transport systems on and off.

In the general case, in which the external environment can be an infinitely large number of states, information on the distribution of the possible states in the environment will play an important role in the synthesis of transport systems.

If the dispersion over the environmental states is small, the first step (with a limited number of transport systems) is to use the proposed algorithm to determine the main transport systems that provide the desired potential and concentrations within the cell. In this case (if there are any severe restrictions on the number of transport systems), the regulatory systems may not be used.

If the dispersion over the states of the environment is large, the synthesis algorithm must be modified such that the regulatory systems of transport are synthesized, at least for the major ions.

4.1.3 Ideal Transport System: Simultaneous Optimization of Robustness and Effectiveness

Previously (see Chap. 1), we considered two limiting cases: an effective system of transport (with minimal costs) and a robust system that maintains a constant internal concentration of the ions. We now consider the limiting case in which both of these properties are simultaneously implemented, which can be considered the ideal system of transport.

Imagine that a cell has an unlimited number of different carriers. Although this situation cannot be realized in nature, consider that this limit is of fundamental interest. Suppose that when the concentration of a substance in the environment changes, the transport system also changes to maintain the required internal concentration of the substance with the new external concentration. In the presence of a large number of different transport systems, such a system, which maintains the required concentration value, can always be found. The individual transport systems may differ from each other in either the driving forces ($\Delta\mu_A$) or the stoichiometry of the ion transport.

Above, we considered the case of the regulation of toxic metals in bacteria.

4.1 Optimization Methods Used for Models of Transport

That is the stoichiometry for a given value of $\Delta\mu_A$ can be different for different transport systems. If, however, this value is limited to integer numbers of ions, then, as shown in Fig. 4.1, it might prove difficult to provide a constant internal environment. It is much easier to obtain the desired consistency if a fractional stoichiometry (the ratio of the number of transported ions to the number of ATP molecules) is used. This does not mean that a fractional number of ions can be transferred, but that the ratio of the number of ions transferred to the number of ATP molecules that are used can be fractional. To obtain a fractional stoichiometry ratio, sufficiently large number of transported ions and molecules of ATP should be considered. In this case, their relationship may be, in principle, arbitrary. For example if three ions are transferred into the cell with two molecules of ATP, the stoichiometric ratio is 1.5. However, the simultaneous transfer of a large number of ions as well as the use of a large number of ATP molecules, would lead to complications in the transport system of a cell.

In this case, it is easy to show that the use of a fractional stoichiometry in the limit of a sawtooth dependence (Fig. 4.1) will lead to a horizontal line.

The same reasoning can be used for robust strategies. Thus, the use of fractional stoichiometry yields both robust and effective strategies for the same limit: a straight line, as shown in Fig. 4.1, which corresponds to the normal state. However, this means that the requirements of robustness and efficiency are simultaneously met. If we consider an effective strategy (when at any moment, only one transport system works), we find that the efficiency is unity and the deviation from the norm is arbitrarily small. For robust strategies the deviation from the norm is zero, and the efficiency of two closely related systems of transport will be arbitrarily close to unity.

The analysis of the situation in which many systems of transport transfer a given ion at the same stoichiometry but with different driving forces is similar. In fact, at various driving forces $\Delta\mu_{Ai}$, the switching between transport systems will be similar to that shown Fig. 4.1. If a small difference exists between the nearest available transport systems, the transport system that is arbitrarily close to the ideal (which simultaneously fulfills the conditions of robustness and efficiency) can be received.

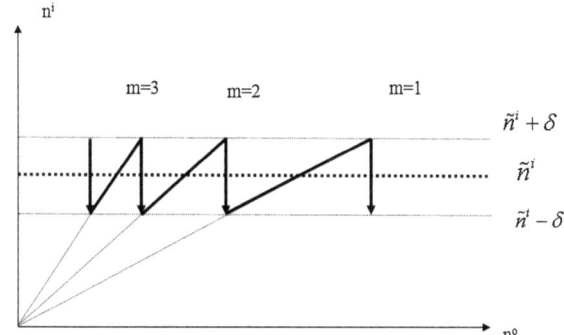

Fig. 4.1 Different stoichiometries for the given value of $\Delta\mu_A$

Fig. 4.2 The sequence of switching between different transport systems

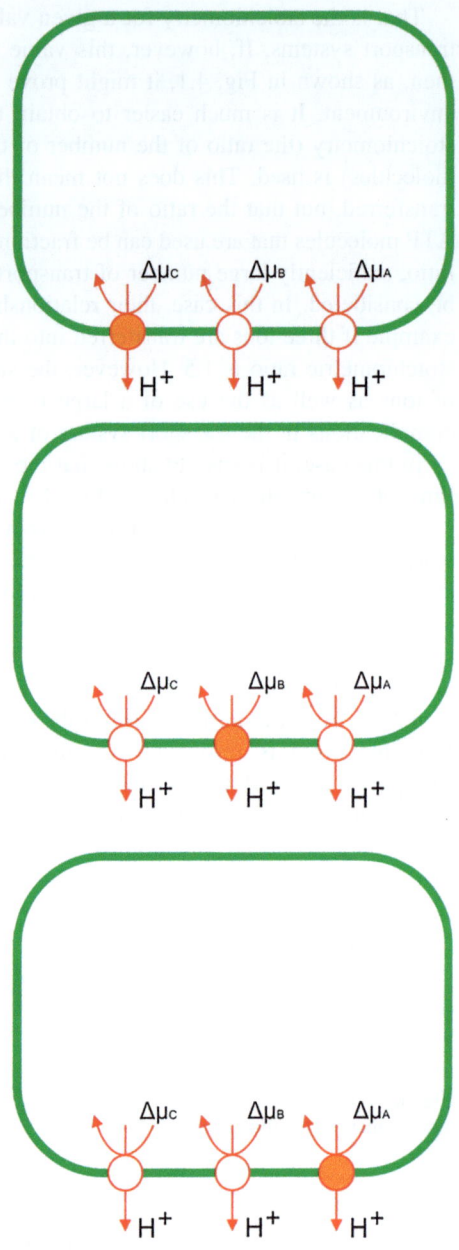

Figure 4.2 illustrates the sequence of switching between different transport systems (effective strategy) during changing of the composition of the environment.

In the case of the system having a robust strategy, the figures will be similar but will involve two adjacent transport systems at all times.

4.1 Optimization Methods Used for Models of Transport

Obviously, the generalization of these cases is the notion of "the driving force attributable to a portable ion". This quantity will eventually characterize the dependence of the internal concentrations on the external ion concentrations.

Note that the presence of charged transferred ions does not introduce any fundamental changes to the algorithm used for the synthesis of an ideal transport system. The only difference in the presence of charged ions would be that the dependence of the internal concentrations on the external ion concentrations is not linear (see Chaps. 2 and 3). However, in this case, the similar transport system that is arbitrarily close to the ideal transport system can be chosen.

4.1.4 Method of the Critical Point

The disadvantage of the ideal system of transport is the need to somehow switch between transport systems. This switching between transport systems also requires modeling. This modeling should take into consideration that any degree of freedom that is associated with the switching between transport systems must somehow be encoded in the genes. Consequently, the relative simplicity (ceteris paribus) is an important feature of this switch.

Consider the simplest version of the switching of transport systems, which was proposed in a previous study (Melkikh and Seleznev 2008). The basis of this switch places a nonlinear dependence of the membrane permeability P for the substance being transported on its external concentration:

$$P(n^0) = P_0 + \alpha(n^0 - \tilde{n}^0) + \frac{\beta}{2}(n^0 - \tilde{n}^0)^2, \qquad (4.7)$$

where P_0 is the constant permeability, α and β are coefficients, and $\tilde{n}^0$ represents the value of the concentration of the substance in the environment, which is near to where the expansion in series is performed.

Equation (4.7) has a fairly simple physical meaning: the permeability can contain both a constant and a variable component, which are proportional to the first and second degree of the concentration of the substance being transported, respectively. This means that the opening or closing of the channel will be a chemical reaction that depends on the fullness of a particular sorption center. It is obvious that one of these centers may contain a single molecule and that the second will contain two of the molecules. This type of membrane permeability control that is caused by the presence of any substance in their sorption centers is widespread. Figure 4.3 illustrates a scheme of the regulation of transport, corresponding to Eq. (4.7).

In accordance with previous research (Melkikh and Seleznev 2008), we consider the simple system of transport for which the regulation described in Eq. (4.7) can be realized. One of these simple systems can be a system that consists of one

Fig. 4.3 The scheme of regulation of transport, corresponding to Eq. 4.7

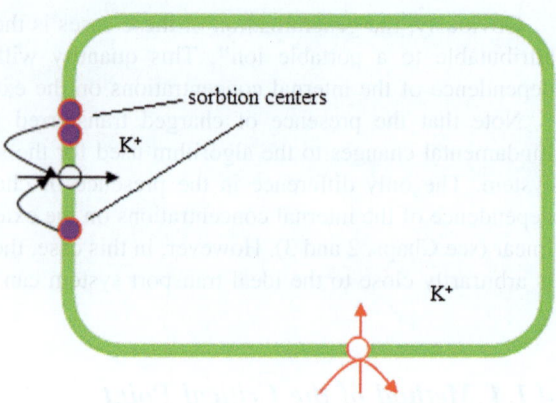

active transporter and one channel with variable permeability. Then, the equation for the flow of the substance being transported will be

$$C(x\exp(\Delta\mu_A) - y) + P(y)(x - y) = 0, \tag{4.8}$$

where C is a constant that characterizes the pump speed, x is the concentration of the substance inside the cell, y is the concentration of the substance in the external environment, and $P(y)$ is the penetrability of the biomembrane to the actively transferred molecules, which can be determined by the following equation:

$$P(y) = P_0 + \alpha(y - y_0) + \frac{\beta}{2}(y - y_0)^2. \tag{4.9}$$

Let x_0 be the internal concentration, the value of which must be maintained at a relatively constant level. Then,

$$P_0 = C\frac{x_0 \exp(\Delta\mu_A) - y_0}{y_0 - x_0}. \tag{4.10}$$

We then choose the coefficients α and β such that the relationship has an inflection point (critical point). At the same time, the following conditions should be satisfied:

$$\frac{\partial}{\partial y}x = 0 \quad \text{and} \quad \frac{\partial^2}{\partial y^2}x = 0.$$

Under these conditions, the dependence of the internal concentrations on the external concentrations becomes critical, as shown Fig. 4.4.

Thus, the method of the critical point uses a nonlinear dependence of the membrane permeability on the concentration of the substance being transported to create a critical point (inflection point) where it is needed. Near this critical point, the dependence of the inner concentration on the outer concentration will be weak.

The proposed method can be extended to the transfer of charged particles and the use of an arbitrary number of transport systems and materials.

4.1 Optimization Methods Used for Models of Transport

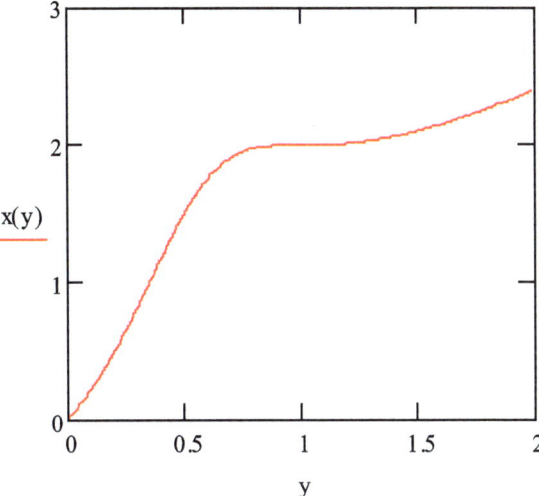

Fig. 4.4 Critical dependence of internal concentration on the external

It may be noticed that the closer each other transport system is to its driving force, the narrower the range of the weak dependence of the internal concentration on the external concentration; the reverse statement is also true. However, the direct losses due to the differences in the transport systems will be less.

The proposed method for the organization of the critical point can also be generalized to an arbitrary number of variables. For example, in the case when an ion is transported by two systems of transport, the concentration inside the cell will depend on the concentrations of the ions outside the cell, which can be used to obtain the concentrations of other ions (for an example, refer to the transport of sodium ions in bacteria, which is described in Chap. 3):

$$n^i_{Na} = n^o_{Na} \frac{P_H P_{Na} + Cn^o_{Na} P_{Na} \exp(\Delta \mu_A) + Cn^o_H P_H \exp(\Delta \mu_A)}{P_{Na} P_H + Cn^o_H P_H + P_{Na} Cn^o_{Na} \exp(\Delta \mu_A) + C^2 n^o_H n^o_{Na} (\exp(\Delta \mu_A) - 1)}. \quad (4.11)$$

After the addition of the permeabilities, P_{Na} and P_H, which depend on the concentrations, to Eq. (4.11), we can obtain the critical point using the same method. The internal concentration of sodium ions will weakly depend on the external concentrations of sodium ions and protons.

The universality of the behavior near the critical point can be manifested: it is always possible to introduce a function of permeability that will provide a critical dependency, regardless of the dependence of the internal concentrations on the outside concentrations. It is important to note that a necessary condition for the organization of the critical point is the presence of "extra" transport systems (active or passive), which can be adjusted accordingly. Above, it was concluded that such "extra" transport systems just play a regulatory role.

It should be noted that a weak dependence of the internal ion concentrations on the external concentrations was also obtained in the simulation of ion transport, (for example, the transfer of potassium ions in plant cells). From the previously derived Eq. (3.53)

$$\varphi = \ln\left(\sqrt{\left(\frac{Z_A n_A}{2 n_{Cl}^o}\right)^2 + \frac{n_K^o}{n_{Cl}^o} + \frac{n_{Na}^o}{n_{Cl}^o}\exp\left(-\frac{\Delta\mu_A}{2}\right)} - \frac{Z_A n_A}{2 n_{Cl}^o}\right)$$

it follows that

$$n_K^i = n_K^o \exp(-\varphi) = \frac{n_K^o}{\sqrt{\left(\frac{Z_A n_A}{2 n_{Cl}^o}\right)^2 + \frac{n_K^o}{n_{Cl}^o} + \frac{n_{Na}^o}{n_{Cl}^o}\exp\left(-\frac{\Delta\mu_A}{2}\right)} - \frac{Z_A n_A}{2 n_{Cl}^o}}.$$

Neglecting the concentration of the sodium ions (because of the small exponent) and expanding the value of

$$\frac{n_K^o}{n_{Cl}^o},$$

in series, we obtain

$$n_K^i = Z_A n_A \left(1 + \frac{n_K^o n_{Cl}^o}{(Z_A n_A)^2} + \cdots\right). \tag{4.12}$$

This formula confirms the earlier conclusion that the internal concentration of potassium ions has a weak dependence on the external concentration. Of course, this dependence is not required to have a critical point, which was used as a workaround in this methodology. Thus, the first term in the series expansion will not include the concentration of potassium ions with high accuracy.

The physical meaning of this relation is clear: the potassium ions account for the vast majority of the cations of plant cells, whereas the majority of anions include the non-penetrating ions, the concentrations of which remain almost unchanged. However, because the condition of neutrality should be upheld in all cases, the concentration of potassium ions must also remain essentially constant when the concentration of potassium ions in the environment changes. This conclusion is also true for other cells, such as bacteria and yeast.

4.1.5 Controllability and Paradox of Ions Transport

Consider the case in which the number of major transport systems is less than the number of transferred ions. Although this situation is usually not realistic for living cells, its study is important for understanding the synthesis of a transport system in artificial cells.

Suppose we have two ions that are transferred through a system of active transport. In addition, we note that both of these ions (for definiteness, let it be the ions

4.1 Optimization Methods Used for Models of Transport

Na$^+$ and H$^+$ ions) can be transferred passively. We then consider the case of transport without electric potential. Then, the flow of ions in the steady state can be written:

$$P_H(n_H^i - n_H^o) + C(n_H^i n_{Na}^o \exp(\Delta\mu_A) - n_H^o n_{Na}^i) = 0 \tag{4.13}$$

and

$$P_{Na}(n_{Na}^i - n_{Na}^o) + C(n_{Na}^i n_H^o - n_{Na}^o n_H^i \exp(\Delta\mu_A)) = 0. \tag{4.14}$$

The solution to the system of Eqs. (4.13 and 4.14) can be written as follows:

$$n_{Na}^i = n_{Na}^o \frac{P_H P_{Na} + C n_{Na}^o P_{Na} \exp(\Delta\mu_A) + C n_H^o P_H \exp(\Delta\mu_A)}{P_{Na} P_H + C n_H^o P_H + P_{Na} C n_{Na}^o \exp(\Delta\mu_A) + C^2 n_H^o n_{Na}^o (\exp(\Delta\mu_A) - 1)}. \tag{4.15}$$

Note that if both passive permeabilities are zero, then Eqs. (4.13 and 4.14) coincide and two variables cannot be found. The problem thus becomes unsolvable.

To solve the system of Eqs. (4.13 and 4.14), we consider a time-dependent formulation of the problem. We therefore let the permeability go to zero and write the balance equations for both ions in the form of

$$V\frac{dn_1^i}{dt} = -C(n_1^i \cdot n_2^o \cdot \exp(\Delta\mu_A) - n_1^o \cdot n_2^i) \tag{4.16}$$

and

$$V\frac{dn_2^i}{dt} = C(n_1^i \cdot n_2^o \cdot \exp(\Delta\mu_A) - n_1^o \cdot n_2^i), \tag{4.17}$$

where V—is the cell volume. Adding both equations, we find that the internal concentration of ions is bound together by the relation

$$x + y = const = x_0 + y_0,$$

where the subscript "0" refers to the initial instant of time. Then, expressing one of the concentrations using this relation, we substitute it into the original equation. The result is a time-dependent set of solutions for the concentrations that contain exponential terms. Next, we let time go to infinity to obtain the stationary solutions:

$$x = b\frac{x_0 + y_0}{a \cdot \exp(\Delta\mu_A) + b} \quad \text{and}$$
$$y = (x_0 + y_0)\frac{a \cdot \exp(\Delta\mu_A)}{a \cdot \exp(\Delta\mu_A) + b}. \tag{4.18}$$

As observed from Eq. (4.18), the distinguishing feature of this solution is that the concentrations of both quantities depend on the initial conditions. However, this is an apparent contradiction because such a situation occurs frequently in mechanics. For example, this situation refers to the position of a particle on a horizontal surface (indifferent balance) or communicating vessels when gravity is negligible. In our case, it will be an inertialess equilibrium because inertia is absent in the model of ion transport.

One can generalize this conclusion to the existence of charge (and potential) and to any number of ions and transport systems.

Note that a similar situation arises in control theory. In linear control theory, these exist systems that are not completely controlled, which means that some variables cannot, by any choice of control, lead to a predetermined value (for details on controllability, see Appendix 2). In our case, this means that, in the presence of one transport system, it is impossible for the two ions to control the concentration of one of the ions (for example, by changing the value of $\Delta\mu_A$) at a given concentration.

We now consider the case in which a transport system transports three ions with the corresponding stoichiometric coefficients m, n and k. Let the active flux of the first component be written as

$$J_1 = mC\left((n_1^i)^m (n_2^i)^n (n_3^i)^k \exp(\Delta\mu_A) - (n_1^o)^m (n_2^o)^n (n_3^o)^k\right).$$

The other fluxes have a similar structure:

$$J_2 = nC\left((n_1^i)^m (n_2^i)^n (n_3^i)^k \exp(\Delta\mu_A) - (n_1^o)^m (n_2^o)^n (n_3^o)^k\right)$$

and

$$J_3 = kC\left((n_1^i)^m (n_2^i)^n (n_3^i)^k \exp(\Delta\mu_A) - (n_1^o)^m (n_2^o)^n (n_3^o)^k\right).$$

We then construct the balance equation for each ion:

$$V\frac{dn_1^i}{dt} = -mC\left((n_1^i)^m (n_2^i)^n (n_3^i)^k \exp(\Delta\mu_A) - (n_1^o)^m (n_2^o)^n (n_3^o)^k\right),$$

$$V\frac{dn_2^i}{dt} = -nC\left((n_1^i)^m (n_2^i)^n (n_3^i)^k \exp(\Delta\mu_A) - (n_1^o)^m (n_2^o)^n (n_3^o)^k\right),$$

$$V\frac{dn_3^i}{dt} = -kC\left((n_1^i)^m (n_2^i)^n (n_3^i)^k \exp(\Delta\mu_A) - (n_1^o)^m (n_2^o)^n (n_3^o)^k\right).$$

It is easy to observe that the equations can be transformed as follows:

$$nk\frac{dn_1^i}{dt} = mk\frac{dn_2^i}{dt} = nm\frac{dn_3^i}{dt}.$$

After integration, we observe that the internal variable concentrations are dependent on each other:

$$nn_1^i = mn_2^i + nn_{10}^i - mn_{20}^i \text{ and}$$
$$kn_2^i = nn_3^i + kn_{20}^i - nn_{30}^i$$

In a living cell, the exact degeneracy of the system of equations will not occur, but weak degeneracy (weak symmetry breaking) is always possible. In this case, when any two ions carry the same basic system of transport and the secondary transport systems for each of them have low power, these two ions are strongly correlated with each other. Such a correlation should occur when the environmental concentrations of each of the ions change.

Summarizing the above examples, we can conclude that the strong links between the different ions lead to the fact that the ability to individually control each of them is limited. From a mathematical point of view, the right-hand sides of the non-stationary ion balance equations can be represented as a column vector:

$$f(n_i) = \begin{pmatrix} f_1(n_i) \\ f_2(n_i) \\ \ldots \\ f_n(n_i) \end{pmatrix}.$$

If this column vector has a rank lower than the number of variables, the system is degenerate and at least one variable is not independent. In contrast to linear control theory, this column vector cannot be represented as a linear combination of variables (matrix) because the ion fluxes depend nonlinearly on their concentrations.

4.1.6 Cascades and Networks of the Transport Molecular Machines

Many biological structures can be represented as networks or graphs. In this case, we can use mathematical methods designed for such objects. In particular, we will use a generalized Kirchhoff's law, which is designed for transport problems. However, the serial or parallel connection of the transport of molecular machines is applicable to many technical processes (for example, the separation of substances). Let us consider the network aspects of the transport of substances in the cells.

Obviously, the same transport systems in a cell (e.g., Na–K-ATPases in the cell membrane) can be considered as they are connected in parallel. A series connection occurs when there are compartments within the cell through which the same ions are transported as through the outer membrane. In general, this symmetry is broken because transport systems of various types are non-linearly related. Therefore, we should represent the transport system as a network of interconnected elements.

One of the first studies in which the metabolic network (which includes the transport networks) was presented in the form of graphs was performed Oster and co-authors (Oster et al. 1971). A number of criteria for the analysis of metabolic and transport networks was later proposed (e.g., Price et al. 2002; Beard et al. 2002).

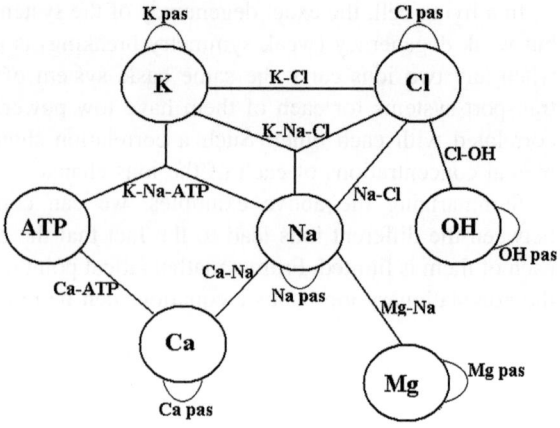

Fig. 4.5 The transport system of a cardiac cell

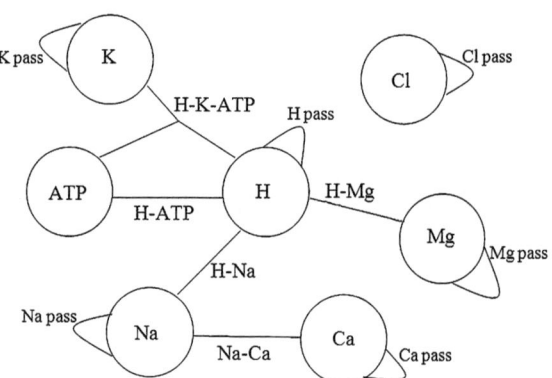

Fig. 4.6 The transport system of archaea

For example, Beard (2002) proposed an analog of Kirchhoff's second law for the analysis of metabolic networks. Other methods include flux balance analysis (FBA) and extreme pathway analysis (ExPA) (Varma and Palsson 1994; Bonarius et al. 1997; Schilling et al. 2000).

In paper (Melkikh and Sutormina 2011), the transport system of cardiac cell was represented as a graph, as shown in Fig. 4.5.

For comparison, we constructed a graph of the transport system of archaea based on the data presented in Chap. 3, which is shown in Fig. 4.6.

In Fig. 4.6, chlorine (as well as all of the other ions) is implicitly linked with the other ions through the potential. This link is not displayed on the graph. Such representation may be useful for understanding the topology of the connections between transport systems.

Because, as was mentioned above, the parallel and serial connections are the limiting cases of the network, it makes sense to analyze these cases separately.

4.1 Optimization Methods Used for Models of Transport

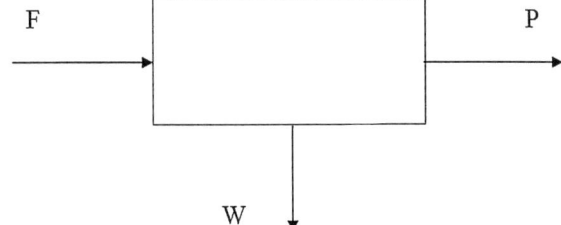

Fig. 4.7 Schematic of a separation element. *F*—feed; *P*—product; *W*—waste

It should be noted that there is an important similarity between the transport of substances in the cells and the process of separating substances in technical systems. For example, the basic unit in the separation of isotopes is a separating element, which, as is the case with a molecular machine, transports material in a cell designed for the transport of a substance. A schematic of the separation element (Palkin 1998) is shown in Fig. 4.7.

The transport system of an ion can be regarded as a special case of this separating element. In fact, each transport system transfers not only one particular ion but transfers other ions with a certain probability (for example, in addition to sodium ions, a transport system has a certain probability of also transferring rubidium ions). This situation generally corresponds to waste products (i.e., generally speaking, unnecessary product).

The similarities and differences between separating elements and the transport systems in cells are shown in Table 4.1.

The difference between the transport of substances in the cell from the cascade separation of substances (except, in the physics of transport, size, etc.) is in the fact that the concentrations of some substances in cells can vary greatly. Therefore, the transport systems in cells are asymmetric, i.e., the individual subsystems that transport different ions are not necessarily constructed in the same way.

We also note an important property of the cascades of separating elements, which is based on the fact that these elements can be (at least in principle) used to build an ideal cascade (e.g., Palkin 1998). The essence of the ideal cascade is that the work that performed for the separation of the mixture in one stage is not lost in the connection between the steps in the cascade. This requires that the mixture streams, that connect the input of any one stage, have the same isotopic composition. In addition, the countercurrent cascade, which only connects with the same concentrations, is ideal, i.e. the entropy of the conjunction of fluxes at the input of each stage does not increase (Cohen 1952).

A similar principle was introduced when considering the efficiency of the cellular transport machinery (Melkikh and Seleznev 2008): the loss of efficiency through the connection of two transport systems is zero only if the systems have the same driving forces. In this case, the concentrations in the steady state are also equal to each other (for example, the various transporters of calcium in cardiac cells, which are described in Chap. 2).

Consider the organization of an artificial cascade in a cell, which was built using the following principle: two adjacent transport systems (which differ little

Table 4.1 Similarities and differences between separating elements and transport systems in cells

Separating element	Pump, exchanger, channel
The separation of substances requires an external power source	The transport of some ions requires ATP energy
Many separating machines are connected in parallel	Many transport systems are connected in parallel
Series-connected separation elements form a cascade	A serial connection of transport systems is possible in compartments
The concentrations of substances change as a result of the separation process	The concentration of substances changes as a result of the transport process
There exists a product-less mode in which there are no input and output fluxes	There exists a mode in which the flow of the substance being transported is absent

from each other) transport the same ion and there is one transport system at the output. Suppose that the number of different transport systems is large. In this case, the next stage of the cascade of transport systems will be smaller by a factor of two. The efficiency of the transport process will be close to unity in this situation because the properties of the neighboring transport systems are similar to each other (see Sect. 4.1.1). Let x be the parameter that characterizes a set of different transport systems. Suppose these systems are ordered such that x increases in the transition from one system to another. The flow at the exit of two transport systems with similar properties is of the following form:

$$J = C_1(\exp(\Delta\mu_1) - 1) + C_2(\exp(\Delta\mu_2) - 1) \text{ and}$$

$$C_2 = C_1 + \frac{\partial C_1}{\partial x}\Delta x, \quad \Delta\mu_2 = \Delta\mu_1 + \frac{\partial \Delta\mu_1}{\partial x}\Delta x,$$

where Δx is the difference of the value of x of two neighboring transport systems. Applying Eq. (4.5), we obtain the following difference in the chemical potential:

$$\Delta\mu = \ln \frac{C_1 \exp(\Delta\mu_1) + \left(C_1 + \frac{\partial C_1}{\partial x}\Delta x\right)\exp\left(\Delta\mu_1 + \frac{\partial \Delta\mu_1}{\partial x}\Delta x\right)}{2C_1 + \frac{\partial C_1}{\partial x}\Delta x}.$$

At small Δx,

$$\Delta\mu = \Delta\mu_1 + \frac{1}{2}\frac{\partial \Delta\mu_1}{\partial x}\Delta x.$$

This means that the difference in the chemical potentials of the output machine is equal to the average value of the input differences. The flow is then converted into the following form:

4.1 Optimization Methods Used for Models of Transport

$$J = C_1(\exp(\Delta\mu_1) - 1)$$
$$+ \left(C_1 + \frac{\partial C_1}{\partial x}\Delta x\right)\left(\exp(\Delta\mu_1)\left(1 + \frac{\partial \Delta\mu_1}{\partial x}\Delta x\right) - 1\right)$$
$$= 2C_1(\exp(\Delta\mu_1) - 1)$$
$$+ \Delta x\left(C_1 \exp(\Delta\mu_1)\frac{\partial \Delta\mu_1}{\partial x} + \frac{\partial C_1}{\partial x}[\exp(\Delta\mu_1) - 1]\right).$$
(4.19)

Thus, the organized of similar transport systems that transport the same ion can convert the flow through each transport system to the next stage. Varying the stoichiometry of the transformation (i.e., the ratio of the number of transport systems at the input to the number of systems at the output), we can either increase or decrease the flow of material through each transport system with almost no loss (the efficiency of this transformation for a small Δx is close to unity).

Nonlinear relationships between the different ion transport systems in cells can be used to maintain the concentration of an ion at a constant level. Various examples of such relationships were described in Chaps. 2 and 3. The presence of strong links between the transport subsystems can result in the fact that a significant change in the concentration of an ion in the environment is more or less evenly distributed among the various ions. As a result, the changes in their intracellular concentrations may be relatively low. This principle can be used in the design of transport systems in artificial cells.

It is essential that, in the nominal mode of operation (norm), these auxiliary transport systems do not work. Basis on the known concentrations, the stoichiometry can be chosen such that the flows are minimal.

A cascade of transport systems can be created to maintain a constant concentration of a substance in a different way.

We assume that the transport of A is organized based on the "critical point" (see Sect. 4.2), which means that there is a non-linear regulator that provides a constant concentration of A inside the cell for a certain range of external concentrations. In the next step, if the substance is transferred to the compartment, its intracellular concentration will have an input parameter for the next regulator, which is organized similarly. By creating a cascade of regulators (see Fig. 4.8), we can achieve a high degree of constancy in the concentration of the substance being transported.

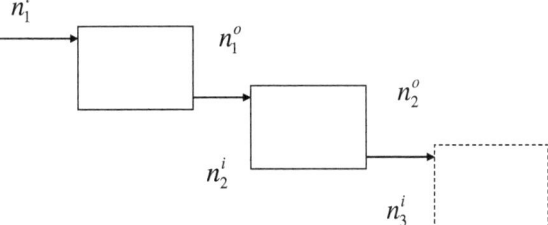

Fig. 4.8 Hierarchical system for the regulation of a transported substance

Thus, we can draw a general conclusion that during the regulation of the transport of one substance, maintained at a constant level at the expense of other (regulatory) substance, concentration of regulatory substance will vary significantly with changes in the environment. These additional degrees of freedom, as well as the process of regulation, will require extra energy.

4.1.7 Regulation of the Pressure in Generalized Cells, Cells in Fresh and Distilled Water, Transport of Water

Consider the cases in which the pressure inside the cell decreases or grows (for example, generalized plant and animal cells).

Let the transport system of a generalized animal cell (Case 1) have the following composition: Na–K-ATPase, passive transport of potassium, passive transport of chlorine ions and non-penetrating ions.

Similarly, the generalized plant cell (Case 2) has the following composition: H-ATPase, Na–H-exchanger, passive transport of potassium and non-penetrating ions.

For Case 1, we have (see Chap. 2)

$$n_{Na}^i = n_{Na}^o \cdot \exp\left(-\varphi - \frac{\Delta\mu_A}{3}\right), \quad n_{Cl}^i = n_{Cl}^o \exp(\varphi e),$$

$$n_K^i = n_K^o \exp(-\varphi) \text{ and}$$

$$\varphi = \ln\left(\sqrt{\left(\frac{Z_A n_A}{2n_{Cl}^o}\right)^2 + \left(\frac{n_{Na}^o}{n_{Cl}^o}\exp\left(\frac{-\Delta\mu_A}{3}\right) + \frac{n_K^o}{n_{Cl}^o}\right)} - \frac{Z_A n_A}{2n_{Cl}^o}\right).$$

Then, we can write the internal concentrations of the major ions and use these to calculate the pressure:

$$n_K^i = \frac{n_K^o}{\sqrt{\left(\frac{Z_A n_A}{2n_{Cl}^o}\right)^2 + \left(\frac{n_{Na}^o}{n_{Cl}^o}\exp\left(-\frac{\Delta\mu_A}{3}\right) + \frac{n_K^o}{n_{Cl}^o}\right)} - \frac{Z_A n_A}{2n_{Cl}^o}}, \quad (4.20)$$

$$n_{Cl}^i = n_{Cl}^o \left(\sqrt{\left(\frac{Z_A n_A}{2n_{Cl}^o}\right)^2 + \left(\frac{n_{Na}^o}{n_{Cl}^o}\exp\left(-\frac{\Delta\mu_A}{3}\right) + \frac{n_K^o}{n_{Cl}^o}\right)} - \frac{Z_A n_A}{2n_{Cl}^o}\right), \quad (4.21)$$

and

$$n_{Na}^i = \frac{n_{Na}^o \exp\left(-\frac{\Delta\mu_A}{3}\right)}{\sqrt{\left(\frac{Z_A n_A}{2n_{Cl}^o}\right)^2 + \left(\frac{n_{Na}^o}{n_{Cl}^o}\exp\left(-\frac{\Delta\mu_A}{3}\right) + \frac{n_K^o}{n_{Cl}^o}\right)} - \frac{Z_A n_A}{2n_{Cl}^o}}. \quad (4.22)$$

4.1 Optimization Methods Used for Models of Transport

Neglecting the small terms, we obtain

$$n_K^i = \frac{n_K^o}{\sqrt{\left(\frac{Z_A n_A}{2n_{Cl}^i}\right)^2 + \frac{n_K^i}{n_{Cl}^i} - \frac{Z_A n_A}{2n_{Cl}^i}}}, \quad n_{Na}^i = \frac{n_{Na}^o \exp\left(-\frac{\Delta \mu_A}{3}\right)}{\sqrt{\left(\frac{Z_A n_A}{2n_{Cl}^i}\right)^2 + \frac{n_K^i}{n_{Cl}^i} - \frac{Z_A n_A}{2n_{Cl}^i}}},$$

and

$$n_{Cl}^i = n_{Cl}^o \left(\sqrt{\left(\frac{Z_A n_A}{2n_{Cl}^i}\right)^2 + \frac{n_K^i}{n_{Cl}^i}} + \frac{Z_A n_A}{2n_{Cl}^i}\right).$$

We assume that the solution is sufficiently dilute such that the pressure inside the cell can be calculated by the following formula:

$$\frac{p^i}{kT} = n_K^i + n_{Na}^i + n_A + n_{Cl}^i$$

$$\approx \frac{n_K^o}{\sqrt{\left(\frac{Z_A n_A}{2n_{Cl}^o}\right)^2 + \frac{n_K^o}{n_{Cl}^o} - \frac{Z_A n_A}{2n_{Cl}^o}}} + n_A + n_{Cl}^o \left(\sqrt{\left(\frac{Z_A n_A}{2n_{Cl}^o}\right)^2 + \frac{n_K^o}{n_{Cl}^o}} + \frac{Z_A n_A}{2n_{Cl}^o}\right). \quad (4.23)$$

If

$$n_K^o \ll n_{Cl}^o$$

then Eq. (4.23) can be expanded in series. As a result, we obtain the following:

$$\frac{p^i}{kT} = Z_A n_A + n_A + n_{Cl}^o \frac{n_K^o}{Z_A n_A}.$$

Thus, the pressure depends only on the concentrations of the potassium and chloride ions. If the charge of the non-penetrating ions is equal to unity, we obtain

$$\frac{p^i}{kT} = 2n_A + n_{Cl}^o \frac{n_K^o}{n_A}.$$

The outside pressure is then

$$\frac{p^o}{kT} = n_{Cl}^o + n_{Na}^o + n_K^o.$$

The difference between these can be written as

$$\frac{\Delta p}{kT} = \frac{p^i}{kT} - \frac{p^o}{kT} = 2n_A + n_{Cl}^o \frac{n_K^o}{n_A} - n_{Cl}^o - n_{Na}^o - n_K^o.$$

Changing the pressure difference across the membrane will depend on how this change would alter the composition of the external environment. We can consider the following two main options: the addition of sodium and chlorine and the addition of potassium and chlorine. If we consider the first case, we have the following equation:

$$\frac{\Delta p}{kT} = 2n_A + (n_{Cl}^o + \alpha)\frac{n_K^o}{n_A} - n_{Cl}^o - n_{Na}^o - n_K^o - 2\alpha,$$

where α represents the addition of sodium and chlorine to the external environment.

Thus, the addition of sodium and chlorine to the environment will cause the pressure difference between the generalized animal cell and its environment to decrease linearly. In the second case, we obtain

$$\frac{\Delta p}{kT} = 2n_A + (n_{Cl}^o + \beta)\frac{(n_K^o + \beta)}{n_A} - n_{Cl}^o - n_{Na}^o - n_K^o - 2\beta,$$

where β represents the addition of potassium with chlorine to the external environment. Thus, there is a nonlinear dependence that, for a small β, will also be decreasing. For large values of β, it is necessary to use the non-linear Eqs. (4.20–4.22).

Recall that all of the calculations were made by changing the pressure difference in the absence of the regulation of ion transport. In the presence of regulation, the results will vary.

From the structure of Eqs. (4.20–4.22), it is observed that the results for the generalized plant cell will be similar.

Consider a special case in which the environment surrounding the cell contains almost no ions. If we assume that the ratios between the concentrations of sodium, potassium, and chlorine remain the same, then the formula for the generalized animal cell that was obtained previously can be simplified to the following:

$$\varphi = \ln\left(\sqrt{\left(\frac{Z_A n_A}{2n_{Cl}^o}\right)^2 + \left(\frac{n_{Na}^o}{n_{Cl}^o}\right)\exp\left(-\frac{\Delta\mu_A}{3}\right) + \frac{n_K^o}{n_{Cl}^o}} - \frac{Z_A n_A}{2n_{Cl}^o}\right) \quad (4.24)$$

$$\approx \ln\left(\frac{n_K^o}{Z_A n_A}\right).$$

If we assume that the concentrations of the non-penetrating ions remain the same and that the concentration of potassium ions is negligible, we find that the potential is negative and very large. A similar conclusion can be drawn for the generalized plant cell.

What are the consequences for the cell from such a significant increase in potential? First, when a large potential difference exists between the cell and its environment, an electrical breakdown of the membrane may occur. Second, the pressure difference between the cell and its environment will be very high. For the generalized animal cell, this may lead to mechanical deformation or the rupture of the membrane. Note, however, that even distilled water contains ions (H^+ and OH^-), which will limit the potential. Although distilled water does not exist in nature, cells that live in fresh water use these or other methods to prevent the deformation and rupture of their membranes.

Consider a mechanism for the removal of water from cells, which exists in some protozoa. One such mechanism involves a contractile vacuole, which is a membrane organelle that is responsible for the release of excess fluid from the cytoplasm. The contractile vacuole is the most significant part of the complex, which acts as the reservoir and is emptied periodically. The fluid enters the contractile vacuole system of the vesicular or tubular vacuoles. The complex allows the cell to maintain more or less a constant volume to compensate for the constant flow of water through the plasma membrane, which is caused by the high osmotic pressure of the cytoplasm.

The contractile vacuoles are common among freshwater protists but are also found in sea forms. Similar structures are found in the cells of freshwater sponges. The main function of the contractile vacuole is the regulation of the osmotic pressure within the body of the protist. The contractile vacuole also partially fulfills an osmoregulatory and renal function by combining the environmental products of the metabolism with water. The contractile vacuole likely plays a significant role in the process of respiration because the water that penetrates the cytoplasm by osmosis carries dissolved oxygen.

However, note that the mechanisms by which the contractile vacuole transports water are not currently clear.

The question of whether there is active transport of water in cells at the level of the water molecules then arises. From the standpoint of thermodynamics, this process is certainly possible, but the literature on this subject is contradictory. There are, for example, water co-transporters (Zeuthen 1995) for instances in which the water molecule is transferred in combination with other ions. However, are water molecules always transferred with ions, or can they be transported separately?

Most likely, the active transport of water is possible only in instances in which the water is somehow consumed (i.e., it reacts with other substances) in any area of the cell. At the same time, a gradient of the chemical potential of the water molecules will form, which will consequently allow the flow of water from one place to another. Because a multi-cellular organism evaporates water, the flow of water, which can be referred to as the active transport of water, must exist.

We can consider a hypothetical transporter of water and calculate the pressure difference that can be created between the cell and the environment. It is important to note that the pressure will also contribute to the chemical potential. Then, taking the pressure difference into account, the chemical potential can be written as

$$\Delta\mu = kT \ln \frac{n^o}{n^i} + v\Delta p,$$

where v is the value per molecule of water. In kT units, the chemical potential can be rewritten as

$$\Delta\mu = \ln\frac{n^o}{n^i} + \frac{\Delta p}{nkT} = \ln\frac{n^o}{n^i} + \frac{M\Delta p}{\rho RT},$$

where M is the molar mass of water and ρ is the density of water.

We can then estimate the contribution of the second term and discuss under which conditions it needs to be considered:

$$\frac{M\Delta p}{\rho RT} \approx \frac{6 \times 10^{-3}}{8.31} \approx 10^{-3}.$$

This contribution is small (it was calculated assuming a pressure of 1 atm), compared with the value of a few units of kT for many ions. However, if a cell only has a system of water transport, this contribution must be considered because it is unique.

In addition to active transport, the passive transport of water will always exist. According to Gennis (1989), the rate constant for the flow of water is 10^{-6} s^{-1}, which means that a water molecule diffuses through the entire thickness of the membrane in 1 mks. If the transmembrane concentration difference is 0.1 M, the water flow rate should be 10–100 molecules per phospholipid molecule.

Consequently, it is likely that a large pressure difference due to the active transport of water cannot be created.

4.1.8 Transport of Ions with a Lack of Energy and Diffusion of ATP

We now consider the case in which the driving force of ion transport is not given, but can be determined from the balance of the number of particles. This situation may arise when the ATP (or other energy source) is supplied to the cell from the outside. This supply through diffusion has a number of inherent difficulties, which, in turn, will lead to a decrease in the value of $\Delta\mu_A$.

We then consider the quasi-stationary case, in which the time changes of the parameters of the environment are sufficiently large.

For example, the following (3.53) expression was obtained for the resting potential in plant cells:

$$\varphi = \ln\left(\sqrt{\left(\frac{Z_A n_A}{2n^o_{Cl}}\right)^2 + \frac{n^o_K}{n^o_{Cl}} + \frac{n^o_{Na}}{n^o_{Cl}}\exp\left(-\frac{\Delta\mu_A}{2}\right)} - \frac{Z_A n_A}{2n^o_{Cl}}\right). \quad (4.25)$$

If the ATP (or another molecule) is supplied to the cell from the outside, then we can write the following equation for the diffusion of ATP (in spherical coordinates) from the external environment:

$$D \frac{1}{r^2} \frac{d}{dr}\left(r^2 \frac{dn}{dr}\right) = 0.$$

where D is the diffusion coefficient of ATP in the external medium. The solution of this equation has the following form:

$$n = C_1 - \frac{C_2}{r}.$$

If the concentration is set to infinity, we obtain

$$n = n_\infty - \frac{C_2}{r}.$$

To determine the second unknown constant, we define the condition of the equality of flows at the cell membrane:

$$-D4\pi R^2 \frac{dn}{dr} = -D4\pi C_2 = -J_A,$$

where J_A—is the consumption of ATP by the cell.
Then, we have

$$n = n_\infty - \frac{J_A}{4\pi Dr}.$$

Consequently, the difference in the chemical potentials will depend on the spatial coordinates and on the rate of ATP consumption by the cell:

$$\Delta \mu_A = \Delta \mu_{A\infty} + \ln\left(1 - \frac{J_A}{4\pi Drn_\infty}\right).$$

This equation states that if the consumption of ATP by the cell is large, the transport will not have sufficient time to supply the energy and the difference in the chemical potentials of ATP-ADP will decrease. Accordingly, the potential will drop, which will lead to a decrease in the difference of the chemical potentials of the ions. On the cell surface, we have

$$\Delta \mu_A = \Delta \mu_{A\infty} + \ln\left(1 - \frac{J_A}{4\pi DRn_\infty}\right).$$

As shown in Chaps. 2 and 3, the consumption of ATP by the cell is due to the passive transport of ions across the membrane (in its absence, ATP is not consumed if there is only active transport). For large passive fluxes across the cell membrane, it is possible for the cell to experience "energy hunger" because the ATP molecules will not have sufficient time to diffuse.

The next question that will be addressed is the following: is it possible to somehow speed up the supply of ATP to avoid this lack of energy? We will thus imagine a system of active transport of ATP and evaluate its effectiveness.

Let ATP be actively transported and assume that ADP and P are transported passively. We can then write an equation for the concentration of ATP within the cell (without charge):

$$n_A^i = n_A^o \exp(\Delta\mu_{A0}). \tag{4.26}$$

Taking the logarithm of Eq. (4.26), we obtain

$$\Delta\mu_{Ai} = 2\Delta\mu_{A0}.$$

Thus, the difference in the chemical potential of ATP-ADP can be increased at the expense of ATP. There is no contradiction because an energy gain does not occur and the full power is not increased. Obviously, the active transport of ADP and P from the cell will lead to a further increase in the value of $\Delta\mu_A$ within the cell. If the active transport of ATP occurs at the expense of another energy source, such a process would constitute the transformation of one form of energy into another.

4.2 Protocells at the Early Stages of Evolution

Because the transport processes are among the most fundamental processes in living cells, it can be assumed that the transport subsystem originated in the earliest stages of evolution. Another aspect of the transport processes is the transport of the protocells themselves, i.e., their movement in space. It can be assumed that the initial mechanisms of transport were similar to those of cellular movement. In this section, we will discuss the modeling of the transport subsystem of cells in the early stages of evolution as well as the transitional phase of directional cell movement as a whole.

There are two different aspects that need to be considered during this analysis: physical and cybernetic. The physical aspect means that it is necessary to build a consistent statistical model of the process. The cybernetic aspect means that it is necessary to consider the simplest behavioral strategy of cells in the early stages of evolution.

4.2.1 Early Stages of Evolution and Origin of the First Cells

According to modern theory, the initiation of chemical processes on the planet was the starting point of chemical evolution. The major result of the first phase was the integration of simple H, C, N and P atoms into relatively complex organic molecules. The next evolutionary process consisted of the integration of these molecules to form macromolecules. Some molecules, which were formed in the proto-atmosphere, were combined into polypeptides and polynucleotides chains, although these molecules were constantly competing with the breakdown process of hydrolysis.

Two possible basic scenarios can be suggested for the processes that are involved in the next step of evolution. In the first scenario, an isolated protocell (a coacervate droplet or a microsphere) appeared first and subsequently acquired its multiplication capability (Fox 1965, 1980, 1988; Fox and Dose 1972; Oparin 1964). The reverse sequence is adopted in the second scenario: the simplest system capable of producing replicas of itself (a replicator) was formed first and the creation of a membrane to isolate it from the environment occurred later. Networks of polynucleotides and polypeptides, which can be referred to as hypercycles, most likely then appeared (Eigen 1971; Eigen and Schuster 1979). Abiogenic polypeptides could have been decisive for the acceleration of the synthesis and replication processes. The modifications of the initial components due to small errors in the replication process and the subsequent selection of favorable variants played an important role in evolution.

A number of models in the literature have described the evolution of self-reproducing molecules or replicators (see, e.g., Fontanari et al. 2006; Szathmary and Demeter 1987; Szathmary 1992).

The importance of metabolism in primitive molecules has been described previously (Morowitz et al. 2000). Some authors believe that the metabolism rather than the replication capability was the first property of living systems.

The gradual complication of molecular complexes led to the formation of protocells (eobionts), which were isolated from the environment by a membrane. These simplest organisms could perform elementary operations with substances (the energy conversion and the transport).

However, even the simplest organisms that exist today are very complex, which leads to the problem of finding the minimum complexity of an organism or a "minimal cell" (Murtas 2007; Glass et al. 2006). This problem can be reasonably tackled in terms of control theory, which states that control is a systemic function that allows the system to retain its structure. The application of this theory to living systems shows that the control process should ensure both the invariability of the spatial structure and the occurrence of a sequence of changes over time, which is connected with the reproduction process.

From the viewpoint of the vital functions of a cell, one of the most important cellular functions consists of the maintenance of constant intracellular conditions, primarily through the composition of the intracellular substances. Therefore, it is necessary to provide an active transport of ions through the cell membrane that does not significantly depend on casual changes in the environmental composition (i.e., the robustness of the process is ensured). Closed membranes that are composed of water-dissolved phospholipids can also be formed under equilibrium conditions. However, in these conditions, the membrane has an average lifetime of τ_0 and the intracellular and extracellular compositions are the same. In this case, a minimum step for the organization of a "living" cell can be assumed to be the establishment of a control system that would maintain the concentrations of substances on both sides of the membrane at different levels. It can then be expected that the intracellular composition is preserved such that the ratio of the membrane life to failure increases and that the system will begin to increase its concentration in the environment. The geographical proliferation of this system can then be initiated because its lifetime is longer than τ_0. For this process to occur, it suffices that a molecule, which acts as an ion pump that absorbs the sunlight or some other form of energy, is built into the material of the cell membrane. In addition, the spatial spreading requires that the intracellular composition and hence the chemical reactions in the cell depend little on changes in the environment and other extracellular conditions, which are inevitable during the proliferation process.

4.2.2 A Model of the Simplest Transport System in a Minimal Cell

The model of the simplest transport system of a protocell was constructed by Melkikh and Seleznev (2008). Based on this study (Melkikh and Seleznev 2008), let us consider a model of the simplest molecular machine that can transport substances to (or from) a protocell in the presence of an external energy source.

We can conceive that the simplest protocell has a membrane that consists of one pump that transfers some substances from the external solution to the interior of the cell. At the start of evolution, this pump could not be too specific and most likely was unable to recognize some molecules. An immediate analog of such a system among the currently existing systems is the synaptic vesicles in neurons that transport neuromediators (Melkikh and Seleznev 2007), as described briefly in Chap. 2.

We will assume the existence of a permanently maintainable source of energy, which could be presented by sunlight or some energy-rich molecules, such as ATP in contemporary cells. We can then because the formula of ATP is not of principal importance in the model and can thus be replaced by another molecule. Thus, the energy source will be simulated as an ATP molecule that can transform to ADP through hydrolysis. The motive force of this source is the difference of the chemical potentials, or $\Delta\mu_A$. It is important (for efficiency reasons) that the

Fig. 4.9 Protocell with simplest transport system of substances

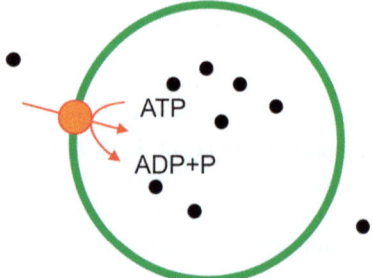

4.2 Protocells at the Early Stages of Evolution

reaction must be catalytic and occur in a transport molecular machine rather than in solution. Figure 4.9 shows a schematic of an efficient transport system in a protocell.

In accordance with the models of transport discussed above, the active flow of a substance to the inside of a vesicle can be written in the following form:

$$J = C(\exp(\Delta\mu_A + \varphi)n^o - n^i). \tag{4.27}$$

We will then assume that the passive flow of this actively transferred substance is small compared to its active flow. Equating the active flow from Eq. (4.27) to zero based on the condition of conservation of the number of particles gives the ratio of the molecule concentrations on the inside to the concentrations on the outside of the protocell:

$$\frac{n^i}{n^o} = \exp(\Delta\mu_A + \varphi). \tag{4.28}$$

For simplicity, let us assume that the prebiotic soup includes only two types of ions: a positive ion, which is actively transferred to the protocell, and a negative ion, which passively penetrates through the cell membrane. Then, the electroneutrality condition should be fulfilled both in the external environment and in the interior of the protocell:

$$n^o_+ = n^o_- \text{ and } n^i_+ = n^i_- \tag{4.29}$$

should be fulfilled both in the external environment and on the inside of the protocell.

The second (negative) component will be distributed inside the protocell in accordance with the Boltzmann law:

$$n^i_- = n^o_- \exp(-\varphi). \tag{4.30}$$

Then, from Eqs. (4.28–4.30), we have

$$\varphi = \frac{\Delta\mu_A}{2} \tag{4.31}$$

and

$$\frac{n^i_+}{n^o_+} = \exp\left(\frac{\Delta\mu_A}{2}\right). \tag{4.32}$$

It can easily be shown from Eqs. (4.31 and 4.32) that the free energy consumed in the transfer of one ion through the membrane equals $\Delta\mu_A$. As shown in Chap. 2, this situation corresponds to the efficiency of unity.

Note that Eq. (4.32) implies that the potential across the membrane does not depend on the ion concentrations in the environment. Therefore, this type of system can be used to maintain the relative constancy of the potential in protocells. If the transferred molecule is uncharged, we have

$$\frac{n^i}{n^o} = \exp(\Delta\mu_A). \quad (4.33)$$

In this case the electrical portion of the free energy is absent, but the energy consumption again equals $\Delta\mu_A$.

In both cases this pump provides a large concentration of the transferred substance inside the protocell (or removes harmful substances from the protocell). For example, if $\Delta\mu_A \approx 20$ kT, we find that the ratio between the intracellular and the extracellular concentrations of the uncharged molecules will be approximately 5×10^8. The free energy of the substance can then be used in further operations (e.g., work, synthesis of other substances, etc.).

One of the simplest cases of transport in the protocell is also the case in which both ions are actively transported. We can then write the following equations:

$$n^i_+ = n^o_+ \exp(\Delta\mu_A + \varphi) \text{ and}$$
$$n^i_- = n^o_- \exp(-\Delta\mu_A - \varphi).$$

It is assumed that the singly charged negative ions are also actively transported into the cell. Then, we obtain

$$n^o_+ \exp(\Delta\mu_A + \varphi) = n^o_- \exp(-\Delta\mu_A - \varphi),$$

$$\varphi = -\Delta\mu_A. \quad (4.34)$$

Apparently, this value of the potential is the maximum (in the absence of nonpenetrating ions) for a given value of $\Delta\mu_A$. Another finding is that a large magnitude of the potential (400 mV) may lead to the electrical breakdown of the cell membranes.

At this stage of the evolution, the high efficiency of the molecular machine is not necessarily the result of a selection process. It might simply be due to a certain structure of molecules that participate in the process of active transport and the subsequent chemical metabolism. This structure can exist even before replication.

In accordance with a previous study (Melkikh and Seleznev 2008), let us consider the problem of robustness of the ion transport system. As shown in Chap. 2, a linear relationship (4.33) does not allow the concentration of the substance being transported to remain constant when the external concentration changes.

4.2.3 A Model of the Simplest System for the Control of Transport Processes in a Cell

Thus, a protocell should have the simplest system for the control of the active transport of ions that reduces the effect of environmental variations.

One of the most important characteristics of the control system is the controller, which is a device that ensures the required behavior of dynamic systems by means

4.2 Protocells at the Early Stages of Evolution

of feedback (Dorf and Bishop 2004). The classical system of feedback control, which was proposed by Wiener, was discussed in Chap. 1.

In accordance with a previous study performed by Melkikh and Seleznev (2008), we will discuss a quasi-stationary controller in a model that describes the control system of transport processes in a minimal cell.

The control device in this system can be either the presence of passive penetrability of the actively transferred ion or one additional systems of active transport. Let us consider these two mechanisms in the analysis of the transport of an uncharged molecule. The controller is based on the method of the critical point, which could be one of the simplest in the early stages of evolution.

In the case in which there are two pumps and only one of the pumps is variable,

$$C_1(x\exp(\Delta\mu_A) - y) + C_2(y)(x\exp(\Delta\mu_B) - y) = 0, \qquad (4.35)$$

which can be rewritten to obtain the intracellular concentration:

$$x = y\frac{C_2(y) + C_1}{C_2(y)\exp(\Delta\mu_B) + C_1\exp(\Delta\mu_A)}.$$

Expanding $C_2(y)$ as a power series with the term $(y-y_0)$ gives

$$C_2(y) = C_0 + \alpha(y - y_0) + \frac{\beta}{2}(y - y_0)^2,$$

where

$$C_0 = C_1\frac{x_0\exp(\Delta\mu_A) - y_0\exp(\Delta\mu_B)}{y_0 - x_0\exp(\Delta\mu_B)}.$$

Equating the derivative $\partial x/\partial y$ to zero gives the expression for α:

$$\alpha = C_1 x_0 \frac{\exp(\Delta\mu_B) - \exp(\Delta\mu_A)}{(x_0\exp(\Delta\mu_B) - y_0)^2}.$$

Similarly, equating the second derivative to zero gives the expression for

$$\beta = \frac{2x_0 C_1}{(y_0 - x_0\exp(\Delta\mu_B))^3}(\exp(\Delta\mu_A) - \exp(\Delta\mu_B)).$$

Figure 4.10 presents the $x(y)$ dependence near the critical point ($x_0 = 2$, $y_0 = 1$) for $\Delta\mu_A = 2$ and $\Delta\mu_B = 2.3$.

The transfer of charged particles is analogous in many respects. If there are two ions (a positive ion, which is actively and passively transferred, and a negative ion, which is only passively transferred), then, instead of Eq. (4.35), we have

$$C(x\exp(\Delta\mu_A + \varphi) - y) + P(y)(x\exp(\varphi) - y) = 0. \qquad (4.36)$$

Fig. 4.10 The dependence between the intracellular and extracellular concentrations

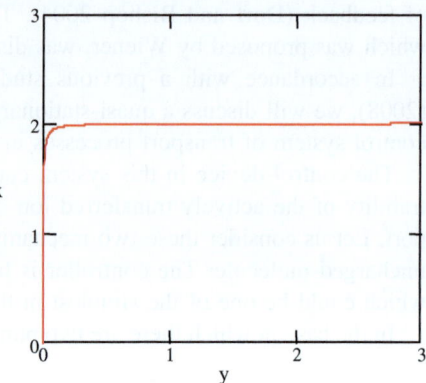

The potential can be expressed in terms of neutrality:

$$x = y \exp(\varphi).$$

Then, we obtain the dependence of the internal concentration on the external:

$$x = y \sqrt{\frac{C + P(y)}{C \exp(\Delta \mu_A) + P(y)}}.$$

Applying a similar method, we can obtain a critical dependency.

The same results can be achieved if we use the same value of $\Delta\mu_A$ but different stoichiometries of the molecule transfer by two active transport systems. In this case, the value of $\Delta\mu_B$ or the stoichiometry constants will present a control instrument that provides an additional method by which the robustness can be increased. A schematic of a protocell with two ion transport systems is shown in Fig. 4.11.

Thus, the variable permeability of the protocell membrane to the substance being transported as well as the variable operational speed of the systems of transport allows for a weak dependence of the internal concentration on the external concentration.

Fig. 4.11 A schematic of a protocell with two ion transport systems

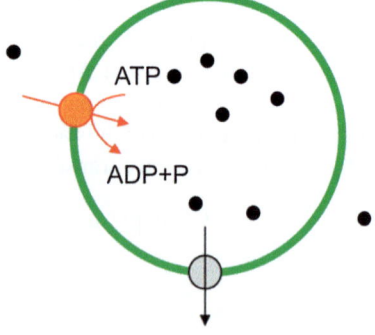

Note that the presence of at least one control system can be considered a definition of life. In fact, all existing control systems are either living systems or systems that subsist on living systems (e.g., technical, economical). Thus, a minimal cell can be considered a system in which substances are transported at the expense of external energy sources and in which the internal concentration of at least one substance is maintained within some limits even if its concentration in the external medium changes considerably. It is this property that distinguishes a minimal cell from molecules and molecular complexes, which cannot be thought of as living systems. The structure of this minimal cell is much simpler than the structures of currently available cells. In other words, the evolution of a cell required a sufficiently long time before the advent of the simplest cells (e.g., archaebacteria) as we understand them today.

4.2.4 Physico-Chemical Models of Cellular Movement

One of the most important properties of living systems, which also arose in the early stages of evolution, was the ability to move directionally.

The current models of simple cells usually do not consider movement. These are limited to the metabolism, the transport of substances and the genetic processes (Murtas 2007; Ganti 2003; Rasmussen et al. 2004; Munteanu and Sole 2006; Melkikh and Seleznev 2008). On the other hand, a number of models have been developed to study the movement of modern cells (Selmeczi et al. 2005; Mora et al. 2009; Pallen and Matzke 2006; Kaneshiro et al. 2001; Thaler and Haimo 1996; McBride 2001; Gracheva and Othmer 2004; Coskun and Coskun 2011). However, these models do not consider the transitional stages.

Prokaryotic and eukaryotic cells move using flagella and cilia. The rotation of the protein structures of flagella leads to both rotary and forward motion (Mora et al. 2009; Pallen and Matzke 2006; Kaneshiro et al. 2001). The flagellum rotation uses the gradient of protons or sodium ions on the cell membrane. The speed of the rotation is approximately 100 Hz. The microscopic mechanism of the transformation of electrochemical energy into movement energy has not yet been studied. It is important to mention that the flagellum of bacteria is a rather complex system and that its mechanism of movement is unclear (a discussion of the mechanisms of flagella evolution is presented in Pallen and Matzke 2006). Because such a complex mechanism could not appear as a whole, it is evident that intermediate stages must exist in which the "motor" structure was much simpler.

At the expense of asymmetric cycles, cilia provoke fluid movement along the surface of a cell. As a rule, large quantities of cilia are present on the surface. Cilia movement is also complex because it involves surface waves (Thaler and Haimo 1996).

There is also a much slower mechanism of cellular movement: the sliding movement of bacteria. This process, however, has not been studied thoroughly (see, for example, McBride 2001).

Because the initial single-celled organisms moved in water, the swimming behavior of such systems is important. A number of models have considered the swimming of bacteria at low Reynolds numbers ("micro-swimming"), which can be obtained with a small body moving in a fluid.

We will now consider the hydrodynamics of micro-swimming (Lauga and Powers 2009).

Swimming at low Reynolds numbers has its own characteristics because any perturbation in the liquid decays rapidly. In this case, the motion of the body in a fluid is described by the stationary Navier–Stokes equations, in which the convective term is neglected:

$$\eta \Delta \vec{u} = \vec{\nabla} p \text{ and } \vec{\nabla} \vec{u} = 0.$$

For example, the Reynolds number for E. coli is in the range of 10^{-4}–10^{-5}, whereas the Reynolds number of some protozoa—is 0.1. Because the rotation of their flexible filaments protozoa reaches a speed of 25–35 mcm/s.

The Scallop theorem is important for the understanding of micro-swimming (Purcell 1997). This theorem states that if a shape change is reversible over time, the movement is absent on average. The simplest example in which a periodic motion allows directional swimming is a traveling wave. An alternative example, in which directional swimming does not exist, is a standing wave. Purcell considered the simplest system that can swim directly, which is referred to as a "Purcell swimmer". Figure 4.12 shows the successive operational stages of this device.

It can be observed that the Purcell swimmer is similar to part of a traveling wave.

In a previous paper (Lauga and Powers 2009), Lauga and Powers considered artificial swimming systems and their optimization. For example, if we define efficiency as the ratio of useful work per unit time to the spent work, it is possible to study the optimization problem. It is shown that the efficiency of flagella is approximately 8 %, whereas the efficiency of living cells is approximately 1 %.

However, there are models of "active Brownian motion" (see, for example, Ebeling et al. 1999; Narumi et al. 2011), in which the motion of the particles, which may not necessarily be considered alive, is modeled. One of the sources of

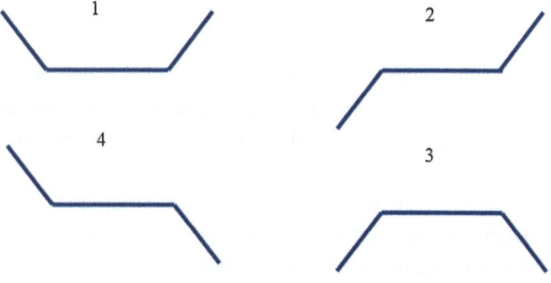

Fig. 4.12 Schematic of Purcell swimmer

4.2 Protocells at the Early Stages of Evolution

energy for the motion of active Brownian particles is the series of specially organized fluctuations in the external environment.

If we consider the first possible mechanisms of cellular movement, the following questions arise. What mechanisms are responsible for the conversion of the stored free energy of chemical reactions into mechanical energy? What could be the simplest mechanisms of movement at the early stages of evolution? Under what conditions does directed movement provide an advantage for natural selection?

A previous article (Melkikh and Chesnokova 2012) described the construction of a model of the simplest form of motion of protocells that could occur in the early stages of evolution, which we will now consider in detail.

Physicochemical Model of Cell Movement

Let us first discuss the possibility of elaboration the model of movement based on physicochemical principles. We will then address the conditions in the biosphere at the early stages of evolution under which the proposed model could work.

We shall consider the model of movement on the basis of a two-level molecule that can be found under non-equilibrium conditions and do work. This model is used for the active transport of ions (see Chap. 1).

We will also consider the following mechanism as the simplest mechanism of movement: a conformon molecule gains energy at the expense of some source and changes its conformation, and, as a result, the energy of the conformon molecule turns into energy for the directed movement of water molecules. According to the law of conservation of momentum, the directed motion must result in the reverse movement of the cell itself. When the energy is transmitted to the water molecules, the conformon returns to its initial state and the process begins again (Fig. 4.13).

The probabilities of finding a conformon molecule in its excited and basic states are

$$f^* = \frac{e^{\Delta\mu - Q}}{1 + e^{\Delta\mu - Q}} \text{ and } f = \frac{1}{1 + e^{\Delta\mu - Q}}.$$

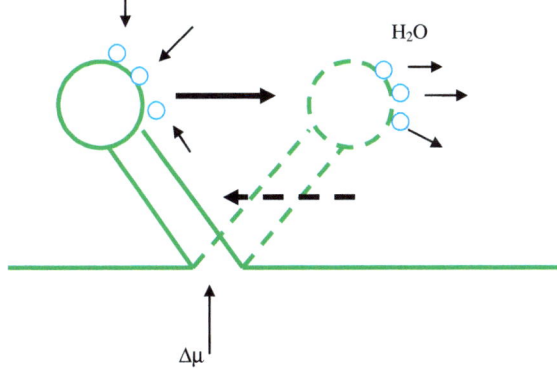

Fig. 4.13 The energy of a conform on turns into energy that is used for the directed movement of water molecules

respectively, where Q is the difference between the energy levels corresponding to the basic and excited states of the conformon and $\Delta\mu$ is the difference of the chemical potentials of the energy source.

When this machine performs lossless work, the difference in the energy levels of the conformon completely transforms into kinetic energy for the movement of water molecules:

$$Q = \frac{m_M v_M^2}{2}.$$

Thus, the mass of the moving conformon should be approximately equal to the mass of the transported water molecules. Hence, we can calculate the velocity that is acquired by the water molecules as a result of the transition of the conformon from its excited state to its basic state:

$$v_M = \sqrt{\frac{2Q}{m_M}},$$

where m_M is the mass of the molecules to which energy is given.

It is extremely important to note that this process is reversible and probabilistic. Thus, the reverse process is possible: the capture of water molecules by a conformon in its basic state leads to its transition to its excited state. The rates of the direct and reverse process are:

$$v' \frac{e^{\Delta\mu-Q}}{1+e^{\Delta\mu-Q}} \text{ and } v \frac{1}{1+e^{\Delta\mu-Q}}.$$

Then, the resultant change in the water molecule impulse per second is equal to

$$\begin{aligned}\frac{d}{dt}\langle m_M v_M \rangle &= v' \frac{e^{\Delta\mu-Q}}{1+e^{\Delta\mu-Q}} m_M v_M - v \frac{1}{1+e^{\Delta\mu-Q}} m_M v_M \\ &= m_M v_M \frac{v' e^{\Delta\mu-Q} - v}{1+e^{\Delta\mu-Q}}.\end{aligned} \quad (4.37)$$

If the conformon is in thermodynamic equilibrium with the environment, the average rate of the directed movement must be equal to zero. Thus, we have

$$v = v' e^{-Q}.$$

We can then obtain

$$\frac{d}{dt}\langle m_M v_M \rangle = m_M v_M \frac{v' e^{\Delta\mu-Q} - v' e^{-Q}}{1+e^{\Delta\mu-Q}} = m_M v_M \frac{v' e^{-Q}(e^{\Delta\mu}-1)}{1+e^{\Delta\mu-Q}}. \quad (4.38)$$

When the movement is stationary, the Stokes force must be equal to the pulling force that is created by the change in the particle impulse:

4.2 Protocells at the Early Stages of Evolution

$$\frac{d}{dt}\langle m_M v_M \rangle = 6\pi R_P \eta v_P, \tag{4.39}$$

where R_P is the protocell radius, η is the environment viscosity, and v_P is the average rate of a protocell movement.

Hence the average speed obtained by a protocell is

$$v_P = \frac{v'}{6\pi R_P \eta}\sqrt{2Qm_M}\frac{e^{-Q}(e^{\Delta\mu}-1)}{1+e^{\Delta\mu-Q}}. \tag{4.40}$$

For example, for $v' = 10^5 Hz$ (the maximum frequency of the work performed by the proton ATPases), $R_P = 10^{-6}m$, $\eta = 10^{-3}Pa \times s$, $Q = 10$, $m_M = 10m_{H_2O}$ and $\Delta\mu \gg 1$, we obtain:

$$v_P \approx 10^{-9} m/s. \tag{4.41}$$

If one protocell have 10^3 these transport systems $v_P \approx 10^{-6}$m/s, which means that a protocell travels a distance equal to its size in approximately 1 s. In any case, this movement is much faster than if the movement mechanism is Brownian motion. Such directed movement is relevant in porous media (ice pores, pyrite pores etc.) and in surface thin films of water in which turbulent mixing is absent.

In general, the magnitude of the velocity of a protocell will depend on its shape and the distribution of the "engines" on its surface.

The calculations carried out by Lauga and Powers (2009) on micro-swimming were based on continuum mechanics, which means that the authors neglected the stochastic processes in the motion. However this is not always true, particularly during the early stages of evolution, when directed motion could be due to the change in the conformation of the molecules. Consider the Purcell swimmer from the standpoint of statistical physics of two-level systems. We assume that each joint is a moving conformon that can exist in one of two states. Obviously, each switch of the conformon from one state to another must require an energy source (e.g., ATP). Assume that both states of each conformon are equal. Thus, it is possible to have four combined states of the two conformons (see Fig. 4.13).

For the emergence of directed motion, the switching of the conformons from one state to another needs to occur in a certain order. However, because this process is probabilistic, some of the switches might fail. Such errors can lead to two different processes: the emergence of a standing wave (in this case, there is no directional movement) and movement in the opposite direction.

Then, we can write an equation for an average speed of protocells, <v>:

$$\begin{aligned}\langle v \rangle &= u\left[\left(\frac{\exp(\Delta\mu_A)}{1+\exp(\Delta\mu_A)}\right)^2 - \left(\frac{1}{1+\exp(\Delta\mu_A)}\right)^2\right] \\ &= u\left(\frac{\exp(\Delta\mu_A)-1}{\exp(\Delta\mu_A)+1}\right).\end{aligned} \tag{4.42}$$

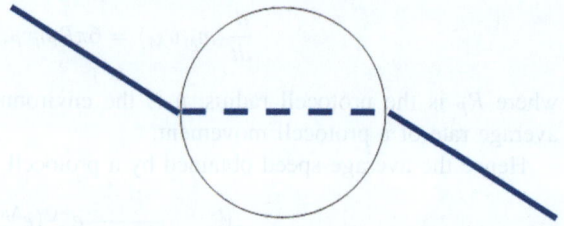

Fig. 4.14 Schematic of protocell with Purcell swimmer

The first term in brackets in the right-hand side of Eq. (4.42) is responsible for the proper implementation of the process of movement, whereas the second term represents the wrong movement (movement in the opposite direction).

In this case, the variable u represents the speed of a Purcell swimmer in a continuous medium and in the absence of reversibility for the conversion of energy from chemical into mechanical form. We observe that, when $\Delta\mu_A \gg 1$, the mean velocity tends to u. However, at $\Delta\mu_A \ll 1$, we obtain

$$\langle v \rangle = u \left(\frac{\Delta\mu_A}{2} \right) \ll u. \tag{4.43}$$

The last inequality is understandable because the average velocity of motion should be zero at zero force (at equilibrium).

It is also important to consider the possibility of the existence of such an engine in the early stages of evolution. Assume that the protocells behave as Purcell swimmers based on changes in their molecular conformations (Fig. 4.14).

From its implementation, it should be obvious that the change in the conformation of the two molecules occur in concert. This can be accomplished either through a direct mechanical connection between the conformons (for example, elastic) or through the transfer of chemical signals inside the protocell.

However, we have not yet addressed the problem of orientation and control in this system. One can show that there is no mechanism that would return the system to its original trajectory after a small deviation from the original trajectory occurs. This statement means that active Brownian motion instead of directed swimming might be responsible for the movement. The presence of receptors and additional "motors" is obviously necessary to arrange a directional movement in the environment. This concept will be discussed in more detail in Sect. 4.2.5.

Consider the problem of movement more broadly with a focus on artificial cells and on alternative forms of life. Let the surface of spherical cells (live or artificial) have a sufficiently large number of elementary "motors". The set of molecular machines can be determined by their distribution on the surface and the direction of the vectors (assuming that these are all equal). We denote $\vec{f}$ the force acting per unit surface area of the moving protocells. Then, the quantity

$$\int_{(S)} \vec{f} dS$$

4.2 Protocells at the Early Stages of Evolution

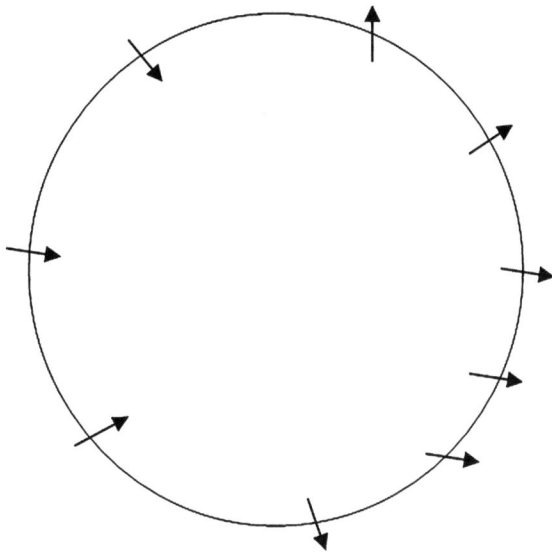

Fig. 4.15 The implementation of more complex forms of motion

is the force acting on the center of mass of the body, and

$$\int\limits_{(S)} \left[\vec{R} \times \vec{f} \right] dS$$

is the total moment of force that will lead to cell rotation.

The implementation of more complex forms of motion is obviously connected with the distribution of "engines" (i.e., the magnitude of $\vec{f}$) on the cell surface (Fig. 4.15).

By creating a set of distribution "engines" on the cell surface and coordinated control, we can achieve the movement of the cell at a given point in time and at a certain speed. In nanomedicine, this type of artificial cells can be considered containers of certain substances. For example, in a previous study (Zhang et al. 2009), Zhang et al. considered the possibility of the creation of artificial flagella.

4.2.5 Sunlight as a Possible Source of Energy for Movement

The model described earlier assumes that the source of energy for a protocell is a chemical that is distributed in the space. This assumption is related to the fact that the most ancient known cells obtain energy from hydrogen: methanogenic bacteria reduce carbon dioxide to methane using molecular hydrogen. However, recent investigations (Gomez-Consarnau et al. 2007; Sabehi et al. 2005) have shown that different variants of photosynthesis (perhaps optional) are widely spread.

Therefore, photosynthesis might play a significant role in the early stages of evolution. For example, in a previous article (Rasmussen et al. 2004) light was considered the energy source of protocells.

In accordance with a previous study performed by Melkikh and Chesnokova (2012), consider a model of the conversion of the energy of photons into the energy of mechanical motion.

According to the thermodynamics of photosynthesis, this process can proceed if the temperature of the incident radiation is higher than the temperature of the environment. In another study (Beatty et al. 2005), the occurrence of photosynthesis in geothermally illuminated environments was proposed. This possibility was indirectly confirmed by a recent discovery of a new type of chlorophyll in cyanobacteria (Chen et al. 2010), which is capable of absorbing red and near-infrared light.

The movement of protocells toward greater illumination can be considered a variance of phototaxis, which many organisms and even organelles (for example, chloroplasts) are known to possess (Suetsugu and Wada 2007; Suetsugu et al. 2010). This type of movement in the presence of non-uniform (variable) illumination could be used as an additional resource for survival.

For example, cyanobacteria perform continuous twenty-four-hour, up-and-down movements in a basin at the expense of the accumulated gases and the buoyancy force that develops (Whitton and Potts 2002). This relatively simple mechanism could also exist at earlier stages of evolution and should be discussed separately.

Let us mention, however, that the mechanisms of directed movement of cells that are related to the direct conversion of light energy into movement energy are unknown. It is possible that these mechanisms existed in the past under definite conditions.

As was obtained in a previous study (Melkikh et al. 2010) that considered the transformation of energy in thylakoids (see also Sect. 3.8),

$$\Delta\mu^{max} \approx h\nu.$$

This two-level system that is destabilized at the expense of sun photons is able to do work on both the synthesis of new protocells and the movement of the cell. In this regard, photons do not differ from any other molecules at the expense of which any work is performed (for example, ATP).

Because the condition $\Delta\mu^{max} \approx h\nu \gg kT$ is met with sunlight, the evaluation for the movement speed shown in Eq. (4.41) can also be applied in this case.

In a previous paper (Melkikh and Chesnokova 2012), a number of evaluations were made from the described protocell movement speed under the assumption that all of the sunlight energy is only used for movement. Taking into account that the power of the sun radiation incident on 1 m^2 of a surface is approximately 10^3 and that only $\chi \sim 1\ \%$ of the incident light is converted into mechanical energy (which is approximately the efficiency of photosynthesis), the useful output incident to a protocell with a radius of 1 micron will be

4.2 Protocells at the Early Stages of Evolution

$$N \approx \pi R^2 <W> \chi \approx 10^{-11} \text{ W}.$$

Equating this output to the power of the friction force, we have

$$v = \left(\frac{R<W>\chi}{6\eta}\right)^{1/2} \approx 10^{-1} \text{m/s}. \qquad (4.44)$$

This speed is certainly higher than the speed of the movement of the microorganism (for example, the movement obtained through the use of flagella is approximately 200 μ/s). This is the upper limit of the movement speed at the expense of sunlight. However, all of the sun energy cannot be used only for movement; a considerable part should be used for the support of other cellular functions. However, it is then necessary to analyze the following question: will a movable or an immovable cell win the competition?

4.2.6 The Energy Balance in Protocells

Let all of the reactions in a protocell proceed at the expense of the free energy of substance A, which is the energy "currency" of the cell. Then, let the reaction, at the expense of which a protocell divides, be the decomposition of substance A into B and C:

$$A \rightleftarrows B + C. \qquad (4.45)$$

The rate of the correspondent decomposition reaction is then

$$J = k_\uparrow n_B n_C (\exp(\Delta \mu_A) - 1),$$

where $\Delta \mu_A = \ln \frac{n_A n_{B0} n_{C0}}{n_B n_C n_{A0}}$ is the affinity of the chemical reaction shown in Eq. (4.45).

In accordance with a previous study (Melkikh and Chesnokova 2012), let us write the equation for the balance of substance A:

$$V \frac{dn_A^i}{dt} = P(n_A^o - n_A^i) - J_D - J_I - J_M, \qquad (4.46)$$

where P is the permeability of the protocell membrane for substance A, n_A^o and n_A^i indicate the concentrations of substance A in the environment and within the protocell, respectively, J_D is the flux of substance A used for the division of the protocell, J_I is the flux of substance A that is used for the work of receptors and the further processing of their signals, J_M is the flux of substance A that is used for protocell movement, and V is the volume of the protocell (Fig. 4.16).

The equation of protocell movement can be written as

Fig. 4.16 Schematic of the energy flows in a protocell

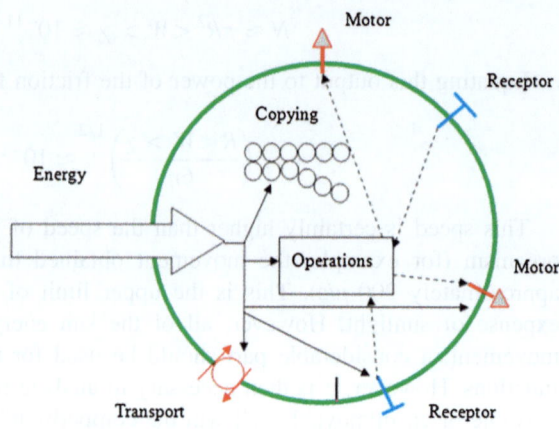

$$\frac{d\vec{p}}{dt} = -\frac{6\pi\eta R}{m}\vec{p} + \vec{F}(I) + \vec{\Phi}, \qquad (4.47)$$

where $\vec{F}(I)$ is the driving force of the "motors", $\vec{\Phi}$ is the force of the random impacts of molecules, and I represents the environmental information that is obtained by the receptors of the protocell. Equation (4.47) is the Langevin equation for controlled Brownian motion. Similar equations have been obtained in the theory of active Brownian motion (see, e.g., Ebeling et al. 1999).

4.2.7 The Problem of Control and Reception of Information: Strategies Used by Protocells for Directed Motion

As mentioned above, the orientation of the protocells in space is crucial to the implementation of directional movement.

In previous papers (Alouges et al. 2008, 2009, 2011), the problem of microswimming in terms of control theory and optimization has been investigated. For example, Alouges (2008) considered a swimming system that consisted of three inter-connected spheres and showed that the system is globally controllable. Because the system of equations of motion at low Reynolds numbers is linear, well-established methods of linear control theory can be applied in the analysis, which showed that local controllability is a necessary condition for swimming.

In this case, it is implicitly assumed that the control system has all of the necessary information about the environment. However, in real microscopic systems, this information is almost always incomplete because a moving body contains a limited number of receptors.

Melkikh and Chesnokova (2012) explored the possibility of driving the system of a protocell, which should have at least two modes of work: orientation changes in space and straight-line movement in the direction chosen.

The minimal movement system that is capable of providing both of these modes of work can have two different organizations:

1. Two symmetrical "motors". The switching-on of only one of these motors leads to an orientation change of the protocell in space. However, the alternate switching-on of both motors provides progressive motion. This behavior can be illustrated by the concerted movements of cilia, but it is a more complicated system that evolved from a simpler one.
2. One "motor" capable of working in both modes (flagellum). In this case, one motor is sufficient, but the system should be more complicated. One motor must provide both modes of work, which involves the changing of the inclination of the developed force to the body.

Three parameters are necessary to define the orientation of a protocell in three-dimensional space. In terms of aerodynamics, these are roll, tangage, and yaw. To provide the possibility to reach any point in space, the tangage and yaw parameters are sufficient. The roll characterizes only the rotation about the axis along the body. The ability of this characteristic control is of principal significance in some processes, e.g., in *Euglena viridis* phototaxis, the cell alternatively turns its lower side with photoreceptors and its upper side without them. However, the roll is not significant if the receptor locations are axisymmetric or if the irritant is distributed uniformly in space.

The precise number and location of the "motors" depends on the specific shape of a body, particularly on its symmetry, and on the properties of the medium through which it moves. Nevertheless, we can provide some generalities concerning the number and location of the "motors".

One motor that creates a momentum about an axis is sufficient to provide the possibility of rotation about the axis. Similarly, to allow rotation around 2 (or 3) axes 2 (or 3) "motors" are required.

There are two ways to provide the straight-line motion of a body: through the addition of another motor (its action force passes through the center of mass) or though the arrangement of the "rotational" motors such that the resultant force asses through the center of mass.

Thus, the minimum number of motors that is required for movement in three-dimensional space is three. If the body shape is complex (asymmetrical), additional motors might be required.

We now turn to the question about the cost of reception. These costs were included in the overall energy balance of a protocell (4.46). Two questions arise. How important is the reception of information for the protocell? In which cases is the cost of reception not significant?

Reception is a very important element of the transport system of a cell because certain transport systems are turned on and off due to the receipt of a signal that indicates a change in the environment. If the received signal is incorrect, then the work of such a system would be ineffective. However, the elimination of errors in the reception itself requires a certain cost. In this case, a situation may arise in which the receptor error correction may be disadvantageous because the cost of this operation might be too high.

According to the Landauer principle (Landauer 1961), the minimal energy consumption that is necessary to obtain one bit of information by a system is $kT \ln 2$. This minimal work should be performed in the infinitely slow transfer of a system into a definite state. In real systems, the switch of states is executed in a finite amount of time, which increases the energy consumption. The energy consumption is also typical for receptors that trace the environment conditions.

Let a receptor of any substance be found in one of only two states: 0 and 1. In the language of information theory, "correct reception" occurs when a receptor is converted into state 1 when it perceives a molecule of the substance. The reverse conversion indicates "wrong reception".

The probability that the receptor is in state 0 can be written as

$$p_T = \frac{e^{\frac{\Delta \mu_A}{kT}}}{1 + e^{\frac{\Delta \mu_A}{kT}}}$$

because substance A is the "single currency" of the protocell and is also used for the reception. The probability of the reverse transfer is

$$p_F = 1 - p_T = \frac{1}{1 + e^{\frac{\Delta \mu_A}{kT}}}.$$

If the expenditures for elementary operation are sufficiently large ($\Delta \mu_A \gg kT$), the probability of correct reception tends to unity, whereas the probability of wrong reception tends to zero. However, in this case, every bit of information will cost approximately $\Delta \mu_A$.

In which situations is the cost of information processing high? In a constantly changing environment and in an environment with sufficiently large gradients, the cell must constantly receive new information about the environment. Therefore, the energy consumption for the reception and processing of this information may be an important limiting factor in these environments.

In accordance with a previous study (Melkikh and Chesnokova 2012), we now consider the balance of matter and energy in a population of protocells. Consider the division of a protocell at the expense of the free energy of substance A that is released as a result of a chemical reaction. The protocell creation can then be coupled with the chemical reaction shown in Eq. (4.45). In turn, the copy synthesis can be presented as follows:

$$\Omega + m\omega + mA \rightleftarrows 2\Omega + mB + mC.$$

where the protocell is designated as Ω and ω is the low molecular weight substances that a protocell it can be synthesized from. Assume that one molecule of

4.2 Protocells at the Early Stages of Evolution

substance A is used to add one molecule of the low-molecular substance ω to the protocell.

According to Melkikh and Chesnokova (2012), the change in the number of protocells per unit time and unit volume can be written as

$$\frac{dn}{dt} = J_\Omega = k_{\Omega\uparrow} n_B^m n_C^m n^2 \left(\exp\left(\frac{m\Delta\mu_A + \Delta\mu_\omega}{kT} \right) - 1 \right). \quad (4.48)$$

Equation (4.48) shows that the driving force of the reaction of the creation of a protocell copy is the total difference of the chemical potentials of the used substance (energy) and material.

In a uniform case, the balance of substance A in the environment should take into account its consumption by a cell both for the creation of the copy and for motion and reception:

$$\frac{dn_A^o}{dt} = nV \frac{dn_A^i}{dt} + q_A,$$

where q_A is the external source of substance A at the expense of which the total population of the protocells exists.

Table 4.2 presents the basic strategies of movable and immovable (Brownian) protocells, which are based on a previous study by Melkikh and Chesnokova (2012).

If there are one or several sources of food, movable protocells could create very simple ecosystems. For example, ecosystems that contain only one living species have recently been discovered deep within Earth (Chivian et al. 2008).

Melkikh and Chesnokova (2012) suggested that the described simplest mechanism of movement could be the predecessor of all of the known forms of directed motion of microorganisms: flagella, cilia, and pili. According to Ferrer et al. (2007), life could be born in microcavities of pyrite crystals. The movement of protocells in the microcavities of other minerals is also possible. In this case, the orientation and directed movement of the protocell are important because these allow the protocell to obtain an additional source of energy or to reach a place with different conditions, such as a different pH level and a different amount of salinity.

Thus, we can provide a special period in the evolution of life in which the directed motion of protocell played a significant role.

Table 4.2 Basic strategies of movable and immovable protocells

Reception	Active movement Exists	Does not exist
Exists	Strategy can win in a changeable environment and in an environment with opposite gradients of matter and energy	Strategy will lose to the immovable cells
Does not exist	Strategy will lose to the immovable cells. However, if it has memory and partial reception, it may win in a changeable environment	Brownian (immovable) cell

The sequence of evolutionary events can be represented as follows:

metabolic networks, microspheres, replicators, RNA-world → replicators in microspheres → MINIMAL MOVABLE CELL → prokaryotes → eukaryotes.

4.3 The Transport of Large Molecules in Living and Artificial Cells

The question of the fundamental basis of the high efficiency of transport of substances is one of the important unresolved issues that limit our understanding of the mechanisms of transport processes in cells. The proteins that carry ions (e.g., ion pumps, exchangers) have a large number of possible degrees of freedom. However, only a very small portion of this amount is involved in the efficient transformation of energy during the transport of ions.

The same problem arises with the transfer of large molecules across the cell membrane and intracellular compartments. Macromolecules (in this study, these are proteins) should be delivered to a specific location within the cell and act as carriers of information, enzymes, chemical reactions, building materials, etc. Often, this information is transferred by macromolecules within the cell. It is clear that errors in the delivery of macromolecules to a specific place, even those associated with a change in their spatial configurations, will continue to accumulate due to the interactions between these molecules. Such a "catastrophe of errors" might cause the cell to simply stop working. Because we know that this does not happen in nature, it is important to understand the causes of this phenomenon.

Obviously, both of these questions are crucial for the creation of artificial cells.

We first define the term "large molecule" in this context. Because this term refers to the complexity of the molecule, "large" molecules are molecules with a large number of possible conformations (degrees of freedom).

The paradox of the transport of large molecules was formulated by Melkikh and Seleznev (2012) and can be illustrated through autophagosomes, which remove "cell debris", such as misfolded proteins. However, the problem of determining the misfolded protein is NP-hard (i.e., its solution requires an exponentially large number of steps). In fact, according to Berezovskii and Trifonov (2002a), the total number of different conformations of a protein with 150 domains, each of which can be in three states, is approximately

$$3^{150} \approx 10^{72}. \qquad (4.49)$$

The enumeration of these options, according to Berezovskii and Trifonov (2002a) would require approximately 10^{48} years, which is unacceptable for the cell and does not correspond to the experimental data.

Thus, the folding of proteins is closely related to the high efficiency of the transport of transport of large molecules. This connection is because there are a large number of variants in the conformations of molecules (mostly proteins).

Let us consider the transport of large molecules from the viewpoint of statistical physics. Because protein folding is a close issue, we first present a brief overview.

Although papers dedicated to Levinthal's paradox are numerous, there is no agreement regarding not only the solution of this paradox, but even its essence and existence. Many investigators have addressed the problem of the protein folding rate (Levinthal's paradox). This paradox was originally formulated as a disagreement between the characteristic time of protein folding to obtain the native configuration of a protein (seconds or fractions of a second) and the characteristic time required for a protein to enumerate all of its possible conformational states (Levinthal 1968, 1969; Dill 1985), which has been estimated elsewhere.

The calculations made in previous studies (Berezovsky and Trifonov 2002a; Finkelstein and Ptitsyn 2002) led to the conclusion that the full enumeration of all variants of protein conformations is impossible and a protein folds using a method that is predetermined through an unknown rule of folding that is dependent on the sequence of amino acids. The protein folding process is assumed to be connected with the formation of a loop structure of amino acids, in which the loops are the elementary units of the folding process.

Some investigators believe that there is no paradox and that the solution was proposed by Levinthal himself (Rooman et al. 2002).

Other investigators (Berezovsky et al. 2000; Grosberg and Khokhlov 2010; Flory 1969; Shimada and Yamakawa 1984; De Gennes 1990; Berezovsky et al. 2001) also described the importance of the loop structure of proteins and proposed different theories on the evolution of modern proteins from their primitive loop-like ancestors that performed relatively simple functions.

Finkelstein (Finkelstein and Ptitsyn 2002; Finkelstein 2002; Galzitskaya et al. 2001; Finkelstein and Badretdinov 1997), however, proposed that a stable structure automatically has the property of quick folding.

He performed calculations considering that the entropy of a protein molecule is large in its unfolded conformation, whereas its energy is low in its native conformation. Consequently, the free energies ($F = U - TS$) of these states are nearly equal. Finkelstein assumed that quick folding requires at least one fast path leading to a stable structure. This path should be free of a high barrier ΔF. If it is assumed that the decrease in the entropy at each step is compensated by a drop in the energy, then the free energy barrier will prove to be low. Thus, the folding time, which was calculated (Finkelstein and Ptitsyn 2002) from the formula

$$\tau = \tau_0 \exp\left(\frac{\Delta F}{kT}\right) \qquad (4.50)$$

may prove to be short.

The main drawback of this method of proof is that the proposed formula for the folding rate holds only under conditions of local equilibrium, i.e., when the evolving system can repeatedly occupy both states, which are separated by the barrier ΔF.

Furthermore, it in no way follows that the free energy is compensated at each step. If the compensation really takes place at each folding step, the folding process takes a special path, which has been discussed by many investigators (Levinthal 1968, 1969; Berezovsky and Trifonov 2002a, 2002b; Trifonov and Berezovsky 2003; Kloczkowski and Jernigan 2002; Ittah and Haas 1995; Berezovsky et al. 2000; Berezovsky and Trifonov 2001). The existence of this path is intimately connected with the domain and loop structures of many proteins (Yeats and Orengo 2007; Riechmann and Winter 2006). If a special folding path is not assumed, the free energy profile proposed by Finkelstein (smooth and free of large barriers) does not exist. Because of the profile irregularity, the folding is necessarily obstructed by large barriers.

If the free energy profile is irregular, a protein will not have time to repeatedly occupy the initial and final states and, therefore, Eq. (4.50) cannot be used to estimate the characteristic time of folding (even if the free energies of the initial and native states are equal).

In line with the above reasoning, the approach formulated by Zwanzig et al. (1992) to estimate the characteristic time appears to be more consistent. In his paper (Zwanzig et al. 1992), Zwanzig and co-workers developed a statistical model of protein folding as a special case of the general problem of "the first pass". This representation seems to be more rigorous than other approaches. Specifically, the calculations of the time of the first pass (folding of a protein to its native configuration) through the use of random walks showed that the estimates calculated by Levinthal are correct. However, if a small "decline" of several kT is introduced for locally incorrect configurations, the time from the first pass to the fully correct configuration becomes much shorter and will coincide with the experimentally measured time. It was inferred from these calculations that the Levinthal paradox was solved.

Bai (2003, 2006) relates the solution of the paradox to the presence of a special landscape in the protein structure. For example, it is suggested in some studies that proteins have a special funnel-like energy landscape that promotes quick folding. It is assumed that this power landscape and other properties of proteins [e.g., hidden intermediates (Bai 2003, 2006)] resulted from evolution.

Grosberg notes (Grosberg 2002) that the characteristic length of a chain, in which all states can be enumerated, was overestimated by Berezovsky and Trifonov. According to Finkelstein and Grosberg, the time of folding of long proteins is proportional to $\exp(N^{2/3})$. This value is smaller than the value obtained by Levinthal. It was also observed that the actual mechanism of the evolution of these sequences remains unknown.

Grossberg and Khokhlov (2010), as well as Davies (2004), believe that Levinthal's paradox exists and that it is currently unsolved.

Thus, the survey of the literature concerned with protein folding has led to the following conclusions:

- A sufficiently long protein should have a special structure and a special path of folding to obtain the characteristic folding time, which is in the order of one second.

4.3 The Transport of Large Molecules in Living and Artificial Cells

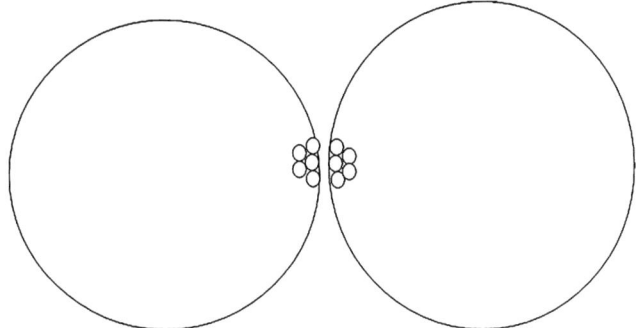

Fig. 4.17 Two interacting polymer globules

- An arbitrary protein (*the one failing the folding time selection*) will not fold quickly. Its characteristic folding time will be exponentially long.

However, the existence of a special way of folding for any sufficiently long proteins has not been proven.

It is clear that the transport of large molecules is only compounded because the interaction between many transported proteins can lead to an even greater number of possible configurations.

However, most biochemical reactions occur according to the principle of "key-lock" or "hand-glove" (Savir and Tlusty 2007).

The essence of these models is that the exact correspondence of reactant molecules exists in a cell to each other, which leads to the high selectivity of the reaction. Given that reaction occurs with high probability, the other variants of the interaction of substances with each other are assumed to be unlikely. In fact, the special path that exists in the space of possible biochemical reactions, leads only to a strictly limited number of reactions and bans all other possible reactions. This result demonstrates the similarity of the biochemical problem of selectivity with the problem of protein folding. However, the statistical justification for the high selectivity of biochemical reactions (specific biochemical pathway) has not yet been addressed in the literature.

We show that the existence of a special method of interaction between the individual domains of a protein and between large molecules is contradictory. Consider two polymer globules that interact with each other and analyze the forces that contribute to this interaction (Fig. 4.17).

There are several types of interactions between atoms of complex molecules. The main interactions include chemical, Van der Waals, and Coulomb forces. The effect of the interaction between atoms (ions) is significant only if

$$U_{12} \gg kT,$$

where U_{12}—is the energy of the interaction between two particles.

In this case, we can consider two types of interactions between two atoms (ions) of different molecules: zone I, in which

$$U_{12} \geq kT$$

and zone II, in which

$$U_{12} \leq kT.$$

According to quantum-chemical calculations, the characteristic size of zone I is 3–5 angstroms, which is approximately equal to the size of the atom.

Thus, if two large molecules are in zone II, then their interaction is too small to affect their movement. In zone II, the random interaction with water molecules must play a major role and the movement of large molecules is random (undirected).

If two large molecules move closer together such that the characteristic energy of interaction between the atoms becomes comparable to kT (zone I), the interaction between the two molecules is significant. At the same time, the selectivity of this interaction may be large, which means that the matching "key" and "lock" of the potential well characterize the interaction as deep.

However, the number of these potential wells will be exponentially large. The majority of the potential wells will be only slightly less deep than the "main" well. If the particle is located in one of the local energy minima, it cannot exactly sense the location of the global minimum. In this situation, the depth of the global minimum is not important. In the theory of stochastic processes, this problem is called "the problem of the first achieving" (Van Kampen 2007), which means that Eq. (4.50) cannot be used. This equation can only be used under local equilibrium conditions, which can be achieved when the particle visits all of the potential wells, the time for which is exponentially large.

We note the following: there may only be a very small number of atoms (the first coordination sphere) in zone one, i.e., in the order of six. However, the number of variants of these atoms is too small compared with the total number of conformations of large molecules, which means that the molecules cannot "see" each other, except for a few surface atoms, during collision.

Thus, the several thousand different protein molecules in the cell must react with all of the other proteins, which will lead to a large number of errors in the interactions of these large molecules. As a consequence, the transport of these molecules will also introduce some errors that will accumulate. The efficiency of the transport of substances in general (including ions) will then also be called into question because any interaction with large molecules will lead to a cascade of "parasitic" biochemical reactions and will ultimately result in a significant reduction in the efficiency of the transport processes.

Examples of the transport of large molecules include the transport of proteins and RNA through the nuclear pores and the transport of proteins in neuronal axons.

An alternative solution of this problem could be the existence of some "long-range" potential for the interactions between molecules, which would have been

4.3 The Transport of Large Molecules in Living and Artificial Cells

Fig. 4.18 Long-range potential

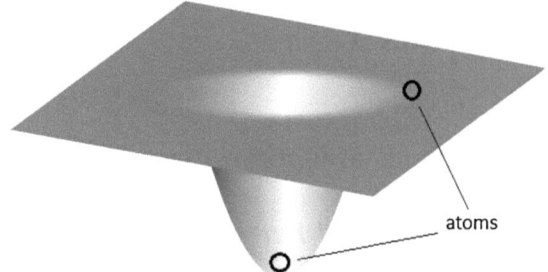

more selective. The characteristic size of this potential (without repulsion) is much larger than the size of the atom (see Fig. 4.18). However, this potential is currently not known.

Thus, the problem of the transport of large molecules is not resolved. However, its solution is crucial for understanding the functioning of living cells and to control the artificial cell. Note also that the accuracy of biochemical reactions is essential for a number of processes, such as DNA computing, artificial storage and the transmission of biological information.

4.4 Conclusion

The models for the transport of substances in cells that are discussed in this book indicate that the transport subsystem of a living cell can be considered optimal. The transport subsystem of a cell is constructed such that it ensures relatively constant internal ion concentrations and minimal transportation energy costs. In the synthesis of artificial cells, the efficiency and robustness of the transport systems are the most relevant features. The above models indicated that these tasks can be solved simultaneously in the limit of a large number of transport systems.

An understanding of the general laws of the optimization of transport subsystems helps create artificial cells with transport systems that are designed for specific tasks, such as the accumulation of valuable ions and compounds and the purification of water from contaminants.

Several possible mechanisms for the directional movement of cells in the early stages of evolution are also discussed. It is shown that the microscopic mechanisms of energy conversion in the transport of matter and in the directed motion of protocells are similar in many respects. It is suggested that the minimal moving cell is a transitional stage between replicators and archaebacteria/cyanobacteria.

Understanding the mechanisms that direct the directional movement of cells will also contribute to the creation of controlled drug carriers.

The problem of the transport of large and complex molecules is also discussed. Currently, this problem does not have a solution. However, an understanding of the mechanism by which the transport of large molecules in the cell occurs with a minimum number of errors is crucial for the creation of artificial cells.

Appendix 1

A.1 Methods of Optimization

A.1.1 Conditional Extremum and Nonlinear Programming

The task for the conditional extremum is formulated as follows:

$$f(x) \to \min, \; g_1(x) = 0, \ldots, g_m(x) = 0$$

where $g_1(x) = 0, \ldots, g_m(x) = 0$—are the constructions that are implied on x.

It can be shown that this problem can be reduced to the optimization of the Lagrangian

$$L(x, \lambda) = f(x) - \sum_{j=1}^{m} \lambda_j g_j(x)$$

on x for a suitable choice of $\lambda_1, \ldots, \lambda_m$ (Lagrange multipliers). As a result, the conditional extremum search is reduced to the solving a system of a equations:

$$\frac{\partial f}{\partial x_k} = \sum_{j=1}^{m} \lambda_j \frac{\partial g_j}{\partial x_k}, \; g_1(x) = 0, \ldots, g_m(x) = 0.$$

A nonlinear programming problem is formulated in a similar manner

$$f(x) \to \min, \; g_i(x) = 0, \; h_j(x) \leq 0$$

where $h_j(x) \leq 0$ are restrictions that are formulated as inequalities. The elements of x that satisfy the constraints of the problem are called the feasible solutions. Through the help of additional variables, the problem can be reduced to

$$f(x) \to \min, \; g(x) = 0, h(x) + z = 0, \; z \geq 0.$$

A special case of nonlinear programming is linear programming, in which the function $f(x)$ has a special form.

A.1.2 Linear Programming

The mathematical formulation of the problem of linear programming can be represented as

Appendix 1

$$\sum_{k=1}^{n} a_{ik}x_k \leq b_i, i = 1, \ldots, m,$$
$$x_k \geq 0, k = 1, \ldots, n,$$
$$w = \sum_{k=1}^{n} c_k x_k \to \max$$

Because the solution of linear programming is achieved in the vertex of the permissible polyhedron—their enumeration ensures success within the finite of steps if solution exists. However, the polyhedron often has an astronomical number of vertices. There are special techniques, such as the simplex algorithm, that allow for a meaningful search.

The classical simplex method (proposed by George Dantzig) implements the idea of a targeted search for solutions. The algorithm jumps over the vertices such that it monotonically increases the target function $\langle c, x \rangle$. During each step, the transition to the next vertex is chosen such that the new value of $\langle c, x \rangle$ is better than the last.

The dual to the classical problem

$$\langle c, x \rangle \to \max, \ Ax \leq b, \ x \geq 0,$$

is the problem

$$\langle b, y \rangle \to \min, \ A^T y \geq c, \ y \geq 0.$$

One of the varieties of linear programming problems is the problem of optimal transportation. This problem is formulated as follows: There are m producers and n consumers of a product, which are located at the nodes of the transportation network. Let x_{ij} denote the amount of the product that is carried out of the ith node to the jth node. In addition, a_i—is the volume of production in the ith node and, b_j—is the total demand in the jth. The natural restrictions on the traffic volume imply:

$$\sum_{j} x_{ij} \leq a_i, \ \sum_{i} x_{ij} \geq b_j.$$

The criterion that is usually used is

$$\sum_{i,j}^{n} c_{ij} x_{ij} \to \min,$$

where c_{ij}—is the cost of the transportation of a unit of goods from the ith to the jth node.

A.1.3 Integer Programming

The situations in real systems commonly lead to problems that are different from the linear programming problems because the desired values of the variables must be integers. The solving of these problems through linear programming is possible, but a situation in which the unknown quantities will be integers in the linear programming solution is a rare exception. In general, the integer programming problem cannot be solved by rounding the resulting values of the linear programming solution to integers. In this case, we need more labor-intensive methods. This problem is NP-hard (i.e., its solution requires an exponentially large number of steps). There are several methods for solving such problems. Two of these are the method of branches and bounds and the method of dynamic programming (see below).

An example of an integer programming problem is the knapsack problem (see Sect. 1.4).

If an exact solution is not required, then the problem is essentially simplified through approximate methods, such as the greedy algorithm, swarm intelligence, and genetic algorithms.

A.1.4 The Packing Problem (Backpack)

The problem of optimal one-dimensional packing (or the backpack problem) is formulated as follows: Suppose we have a backpack with a given carrying capacity and a set of objects of different weights and values. We then want to pack the backpack with the maximum objects such that it closes, which means that the sum of the values and weights of packaged objects must be at maximum and within a given limit, respectively.

There are many variations of this problem that are widely used in practice: the optimal filling of containers, the loading of trucks with weight restrictions, the creation of backups on removable media, and the choice of optimal control in a variety of economic and financial transactions.

There are n objects and c_i and a_i are the cost and weight, respectively, of the i-th object. It is then necessary to select a group of items with the maximum total cost and a limited total weight:

$$\sum_i c_i x_i \to \max, \quad \sum_i a_i x_i \leq W,$$

where x_i can take integer values or the values 1 and 0 (an object either exists or does not exist). This task belongs to the NP-hard class of problems, i.e., its solution requires an exponentially large number of steps that depend on the number of objects.

The methods that can be used to approximate the solution to the backpack problem include genetic algorithms, algorithms of ant colonies, and "greedy" algorithms.

These algorithms are characterized by polynomial complexity but only an approximate solution is obtained. These algorithms are often used for other problems of AI.

The "greedy" algorithm for the knapsack problem is as follows:

The first set of items Q is ordered by descending "specific values" (or the weight of the unit price) items such that

$$\frac{c_1}{a_1} \geq \frac{c_2}{a_2} \geq \ldots \geq \frac{c_n}{a_n}$$

then, starting with an empty set, the sequentially ordered set of objects Q is added to the approximate solution Q' (which is empty at first);

With each successive addition, the algorithm verifies that the next object does not exceed allowable weight of the backpack;

The process is completed by constructing an approximate solution of the knapsack problem.

The exact solution of the knapsack problem can be obtained by other methods, such as the branch and bound method or the dynamic programming method.

A.1.5 Dynamic Programming

Dynamic programming is a method that can be used to solve complex problems by breaking them into simpler subtasks. It is applicable to problems with optimal substructure that include a set of overlapping subproblems with slightly less complexity than the original. In this case, the computation time can be reduced compared to methods that use an exhaustive search algorithm. As a rule, the solution of the problem requires the solving of some of the tasks (subtasks) and the combination of the solutions of the subproblems into a common solution. Often, many of the subtasks are the same. The dynamic programming approach is to solve each subproblem only once, thereby reducing the amount of computation. This is especially useful in cases in which the number of repeated subproblems is exponentially large.

The method of dynamic programming requires the storage of the results of the subtasks that can be used again in the future. Dynamic programming includes the reformulation of a complex problem as a recursive sequence of simpler subproblems.

An optimal substructure in dynamic programming means that the optimal solution of the smaller subproblems can be used to solve the original problem. For example, the shortest path in the graph from vertex A to vertex B can be found by first calculating the shortest path from all of the vertices that are adjacent to A to B and then, taking into account the weights of the edges between A and its adjacent vertices, to choose the best way to B. In general, we can solve the problem, which has an optimal substructure, by performing the following three steps:

1. Split the task into subtasks of smaller size.
2. Find the optimal solution of the subproblems recursively by performing the same three-step algorithm.
3. Use the obtained solutions for the subproblems to construct the solution for the original problem.

The sub-problems are solved by dividing them into even smaller sub-tasks and so on until a trivial case has to be solved in constant time (i.e., immediately). For example, if we need to find the value of n!, it would be a trivial task to determine $1! = 1$ (or $0! = 1$).

Overlapping sub-problems in dynamic programming refer to sub-problems that are used to address a number of larger problems. A striking example is the calculation of the Fibonacci sequence. In this case, a simple recursive approach would be to spend time on the computation of the solution of problems (adding numbers) that have already been solved.

To avoid this exponentially difficult problem, the solution to the already solved sub-problems should be stored. Therefore if the same solution is required in a different task, it can simply be obtained from the memory instead of calculated.

Thus, dynamic programming uses the following properties of the problem:

- overlapping sub-problems,
- an optimal substructure, and
- the ability to memorize the solutions to common subtasks.

The method of dynamic programming was developed by Richard Bellman. The basis of this method is the principle of optimality:

> The optimal strategy has the property that, whatever the initial state and initial decision, the subsequent decision should determine the optimal strategy with respect to the state resulting from the initial decision.

Some classical problems of dynamic programming

- The task of drafting a distance (Levenshtein distance): given two strings, the minimum number of erasures, additions and substitutions of characters that transform one string into another.
- The problem of calculating Fibonacci numbers.
- The problem of selecting a trajectory.
- The making of a consistent decision.
- The packing problem (knapsack): from an unlimited set of objects with "values" and "weights", a set of objects has to be selected to maximize the total value of the limited total weight.

A.1.6 Matrix Games

Matrix games are characterized by two players with conflicting interests. If their interests are not completely opposite, the games are called bimatrix.

The matrix game can be described by the matrix gains

$$\mathbf{A} = \begin{pmatrix} a_{11} & a_{12} & \cdots & a_{1n} \\ a_{21} & a_{22} & \cdots & a_{2n} \\ \cdots & \cdots & \cdots & \cdots \\ a_{m1} & a_{m2} & \cdots & a_{mn} \end{pmatrix},$$

where the elements of the matrix a_{ik} are a winning player (winning player is a_{ik}).

In some cases, the optimization of a system considers the game against nature. In this case, nature represents the second player, who chooses the strategies that are the worst for the first player. It is often convenient to reduce the control problems in the presence of uncertainty (lack of information) to game problems in which the second player is assigned the properties that characterize the random process for which the information is incomplete.

The matrix game often has no solution in pure strategies. In this case, the solution is sought in the form of mixed strategies, i.e., every player uses all of their strategies with a certain probability.

Appendix 2

A.2 Controllability of Linear Control Systems

The behavior of a multidimensional linear control system is described by the equations of state and the output:

$$\dot{x}(t) = Ax(t) + Bu(t), \; x(0) = x_0, \; \text{and}$$
$$y(t) = Cx(t),$$

where x is the n-dimensional vector of the state, u is an r-dimensional vector of the control, t is the time, $t \in [t_0, t_1]$ is the interval of time during which the system functions, y is a k-dimensional vector of the output and A, B, and C and matrices with dimensions of $(n \times n)$, $(n \times r)$, and $(k \times n)$, respectively.

The system is completely controllable if the choice of control action $u(t)$ in the time interval $[t_0, t_1]$ makes it possible to transfer the system from any initial state $x(t_0)$ to an arbitrary predetermined final state $x(t_1)$.

The system is fully controlled by the output if the choice of control action $u(t)$ in the time interval $[t_0, t_1]$ makes it possible to transfer the system from any initial state $x(t_0)$ to a final state with a predetermined arbitrary value of the output $y(t_1)$.

The problem is formulated as follows: given known matrices A, B, C of a system of differential equations, is the system completely controllable?

The criterion of control by the state. For the system to be completely controlled by the state, it is necessary and sufficient that the rank of the controllability by the state

$$W = (B, AB, A^2B, \ldots, A^{n-1}B)$$

be equal to the dimension of the state vector $rang W = n$.

The criterion for the controllability by the output. For the system to be completely controlled by the output, it is necessary and sufficient that the rank of the matrix of the controllability of the output

$$P = (CB, CAB, CA^2B, \ldots, CA^{n-1}B)$$

be equal to the dimension of the vector output $rang P = k$.

References

Alouges F, DeSimone A, Heltai L (2011) Numerical strategies for stroke optimization of axisymmetric microswimmers. Math Models Methods Appl Sci 21(2):361–387

Alouges F, DeSimone A, Lefebvre A (2008) Optimal strokes for low Reynolds number swimmers: an example. J Nonlinear Sci 18(3):277–302

Alouges F, DeSimone A, Lefebvre A (2009) Optimal strokes for axisymmetric microswimmers. Eur Phys J E Soft Matter 28(3):279–284

Bai Y (2003) Hidden intermediates and Levinthal paradox in the folding of small proteins. Biochem Biophys Res Commun 305(4):785–788

Bai Y (2006) Energy barriers, cooperativity, and hidden intermediates in the folding of small proteins. Biochem Biophys Res Commun 340(3):976–983

Beard DA, Liang SD, Qian H (2002) Energy balance for analysis of complex metabolic networks. Biophys J 83(1):79–86

Beatty JT, Overmann J, Lince MT, Manske AK, Lang AS, Blankenship RE, Van Dover CL, Martinson TA, Plumley FG (2005) An obligately photosynthetic bacterial anaerobe from a deep-sea hydrothermal vent. Proc Natl Acad Sci 102(26):9306–9310. doi:10.1073/pnas.0503674102

Berezovsky IN, Grosberg AY, Trifonov EN (2000) Closed loops of nearly standard size: common basic element of protein structure. FEBS Lett 466:283–286

Berezovsky IN, Kirzhner VM, Kirzhner A, Trifonov EN (2001) Protein folding: looping from the hydrophobic nuclei. Proteins 45(4):346–350

Berezovsky IN, Trifonov EN (2001) Van Der Waals locks: loop-n-lock structure of globular proteins. J Mol Biol 307(5):1419–1426

Berezovsky IN, Trifonov EN (2002a) Loop fold structure of proteins: resolution of Levinthal's paradox. J Biomol Struct Dyn 20(1):5–6

References

Berezovsky IN, Trifonov EN (2002b) Back to units of protein folding. J Biomol Struct Dyn 20(3):315–316

Bonarius HPJ, Schmid G, Tramper J (1997) Flux analysis of underdetermined metabolic networks: the quest for the missing constraints. Trends Biotechnol 15:308–314

Chen M, Schliep M, Willows RD, Cai Z-L, Neilan BA, Scheer H (2010) A red-shifted chlorophyll. Science 329(5997):1318–1319

Chivian D, Brodie EL, Alm EJ, Culley DE, Dehal PS, DeSantis TZ, Gihring TM, Lapidus A, Lin L-H, Lowry SR, Moser DP, Richardson PM, Southam G, Wanger G, Pratt LM, Andersen GL, Hazen TC, Brockman FJ, Arkin AP, Onstott TC (2008) Environmental genomics reveals a single-species ecosystem deep within Earth. Science 322(5899):275–278

Cohen K (1952) The Theory of Isotope Separation. McGraw-Hill, New York

Coskun H, Coskun H (2011) Cell physician: reading cell motion. A mathematical diagnostic technique through analysis of single cell motion. Bull Math Biol 73(3):658–682. doi: 10.1007/s11538-010-9580-x

Davies PCW (2004) Does quantum mechanics play a non-trivial role in life? BioSystems 78:69–79

De Gennes PG (1990) Introduction to polymer dynamics. Cambridge University Press, Cambridge

Dill KA (1985) Theory for the folding and stability of globular proteins. Biochemistry 24(6):1501–1509

Dorf RC, Bishop RH (2004) Modern control systems, 10th edn. Prentice-Hall Inc, New Jersey

Ebeling W, Schweitzer F, Tilch B (1999) Active Brownian particles with energy depots modeling animal mobility. BioSystems 49:17–29

Eigen M (1971) Selforganization of matter and the evolution of biological macromolecules. Naturwissenschafen 58(10):465–523

Eigen M, Schuster P (1979) The hypercycle: a principle of natural self-organization. Springer, Berlin

Ferrer M, Golyshina OV, Beloqui A, Golyshin PN, Timms KM (2007) The cellular machinery of Ferroplasma acidiphilum is iron-protein-dominated. Nature 445:91–94

Finkelstein AV (2002) Cunning simplicity of a hierarchical folding. J Biomol Struct Dyn 20(3):311–313

Finkelstein AV, Badretdinov AY (1997) Rate of protein folding near the point of thermodynamic equilibrium between the coil and the most stable chain fold. Fold Des 2(2):115–121

Finkelstein AV, Ptitsyn OB (2002) Protein physics. Academic, Oxford

Flory PJ (1969) Statistical mechanics of chain molecules. Interscience, New York

Fontanari JF, Santos M, Szathmary E (2006) Coexistence and error propagation in pre-biotic vesicle models: a group selection approach. J Theor Biol 239(2):247–256

Fox SW (1965) Simulated natural experiments in spontaneous organization of morphological units from protenoid. In: Fox SW (ed) The origins of prebiological systems and of their molecular matrices. Academic, New York

Fox SW (1980) The origins of behavior in macromolecules and protocells. Comp Biochem Phys B 67(3):423–436

Fox SW (1988) The emergence of life: Darwinian evolution from the inside. Basic Books, New York

Fox SW, Dose K (1972) Molecular evolution and the origin of life. Freeman WH and Co, San Francisco

Galzitskaya OV, Ivankov DN, Finkelstein AV (2001) Folding nuclei in proteins. FEBS Lett 489:113–118

Ganti T (2003) The principles of life. Oxford University Press, Oxford

Gennis RB (1989) Biomembranes. Molecular structure and function. Springer, New York

Glass JI, Assad-Garcia N, Alperovich N, Yooseph S, Lewis MR, Maruf M, Hutchison CA III, Smith HO, Venter JC (2006) Essential genes of a minimal bacterium. Proc Natl Acad Sci 103(2):425–430

Gomez-Consarnau L, Gonzalez JM, Coll-Llado M, Gourdon P, Pascher T, Neutze R, Pedros-Alio C, Pinhassi J (2007) Light stimulates growth of proteorhodopsin-containing marine Flavobacteria. Nature 445:210–213

Gracheva ME, Othmer HG (2004) A continuum model of motility in ameboid cells. Bull Math Biol 66:167–193

Grosberg AY (2002) A few disconnected notes related to Levinthal paradox. J Biomol Struct Dyn 20(3):317–321

Grosberg AY, Khokhlov AR (2010) Giant molecules: here, there, and everywhere, 2nd edn. World Scientific Publishing Company, London

Ittah V, Haas E (1995) Nonlocal interactions stabilize long range loops in the initial folding intermediates of reduced bovine pancreatic trypsin inhibitor. Biochemistry 34(13):4493–4506. doi:10.1021/bi00013a042

Kaneshiro ES, Sanderson MJ, Witman GB (2001) Amoeboid movement, cilia and flagella. In: Sperelakis N (ed) Cell physiology sourcebook, 3rd edn. Academic, San Diego

Kloczkowski A, Jernigan RL (2002) Loop folds in proteins and evolutionary conservation of folding nuclei. J Biomol Struct Dyn 20(3):323–325

Lauga E, Powers TR (2009) The hydrodynamics of swimming microorganisms. Rep Prog Phys 72:096601

Landauer R (1961) Irreversibility and heat generation in the computing process. IBM J Res Devel 5:183–191

Levinthal C (1968) Are there pathways for protein folding? J Chim Phys 65:44–45

Levinthal C (1969) How to fold graciously? Mossbauer spectroscopy in biological systems. In: Debrunner P, Tsibris JCM, Monck E (eds) University of Illinois Press, Urbana, pp 22–24

McBride MJ (2001) Bacterial gliding motility: multiple mechanisms for cell movement over surfaces. Ann Rev Microbiol 55:49–75

Melkikh AV, Seleznev VD (2007) Models of active transport of neurotransmitters in synaptic vesicles. J Theor Biol 248(2):350–353

Melkikh AV, Seleznev VD (2008) Early stages of the evolution of life: a cybernetic approach. Orig Life Evol Biosph 38(4):343–353

Melkikh AV, Seleznev VD, Chesnokova OI (2010) Analytical model of ion transport and conversion of light energy in chloroplasts. J Theor Biol 264(3):702–710

Melkikh AV, Sutormina MI (2011) Algorithms for optimization of the transport system in living and artificial cells. Syst Synth Biol 5:87–96

Melkikh AV, Seleznev VD (2012) Mechanisms and models of the active transport of ions and the transformation of energy in intracellular compartments. Prog Biophys Mol Biol 109(1–2):33–57

Melkikh AV, Chesnokova OI (2012) Origin of the directed movement of protocells in the early stages of the evolution of life. Origins Life Evol B 42(4):317–331

Mora T, Yu H, Sowa Y, Wingreen NS (2009) Steps in the bacterial flagellar motor. PLoS Comput Biol 5(10):e1000540

Morowitz HJ, Kostelnik JD, Yang J, Cody GD (2000) The origin of intermediary metabolism. Proc Natl Acad Sci 97(14):7704–7708

Munteanu A, Sole RV (2006) Phenotypic diversity and chaos in a minimal cell model. J Theor Biol 240(3):434–442

Murtas G (2007) Question 7: construction of a semi-synthetic minimal cell: a model for early living cells. Orig Life Evol Biosp 37(4–5):419–422

Narumi T, Suzuki M, Hidaka Y, Asai T, Kai S (2011) Active Brownian motion in threshold distribution of a Coulomb blockade model. Phys Rev E 84:051137

Oparin AI (1964) Life: its nature, origin and development. Academic, New York

Oster G, Perelson A, Katchalsky A (1971) Network thermodynamics. Nature 234:393–399

Palkin VA (1998) Potential and separative power in separation of binary mixtures of isotopes. At Energ 84(3):196–201

Pallen MJ, Matzke NJ (2006) From the origin of species to the origin of bacterial flagella. Nat Rev Microbiol 4(10):784–790. doi:10.1038/nrmicro1493

Price ND, Famili I, Beard DA, Palsson BO (2002) Extreme pathways and kirchhoff's second law. Biophys J 83(5):2879–2882

Purcell EM (1997) The efficiency of propulsion by a rotating flagellum. Proc Natl Acad Sci 94:11307–11311

Rasmussen SJ, Chen L, Stadler BM, Stadler PF (2004) Proto-organism kinetics: evolutionary dynamics of lipid aggregates with genes and metabolism. Orig Life Evol Biosph 34(1–2): 171–180

Riechmann L, Winter G (2006) Early protein evolution: building domains from ligand-binding polypeptide segments. J Mol Biol 363(2):460–468

Rooman M, Dehouck Y, Kwasigroch JM, Biot C, Gilis D (2002) What is paradoxical about Levinthal paradox? J Biomol Struct Dyn 20(3):327–329

Sabehi G, Loy A, Jung KH, Partha R, Spudich JL, Isaacson T, Hirschberg J, Wagner M, Béjà O (2005) New insights into metabolic properties of marine bacteria encoding proteorhodopsins. PLoS Biol 3(8):e273

Savir Y, Tlusty T (2007) Conformational proofreading: the impact of conformational changes on the specificity of molecular recognition. PLoS ONE 2(5):e468. doi:10.1371/journal.pone.0000468

Selmeczi D, Mosler S, Hagedorn PH, Larsen NB, Flyvbjerg H (2005) Cell motility as persistent random motion: theories from experiments. Bioph J 89(2):912–931

Schilling CH, Letscher D, Palsson BO (2000) Theory for the systemic definition of metabolic pathways and their use in interpreting metabolic function from a pathway-oriented perspective. J Theor Biol 203:229–248

Shimada J, Yamakawa H (1984) Ring-closure probabilities for twisted wormlike chains. Application to DNA. Macromolecules 17(4):689–698

Suetsugu N, Wada M (2007) Chloroplast photorelocation movement mediated by phototropin family proteins in green plants. Biol Chem 388(9):927–935

Suetsugu N, Yamada N, Kagawa T, Yonekura H, Uyeda TQP, Kadota A, Wada M (2010) Two kinesin-like proteins mediate actin-based chloroplast movement in Arabidopsis thaliana. Proc Natl Acad Sci 107(19):8860–8865

Szathmary E (1992) Natural selection and the dynamical coexistence of defective and complementing virus segments. J Theor Biol 157(3):383–406

Szathmary E, Demeter L (1987) Group selection of early replicators and the origin of life. J Teor Biol 128(4):463–486

Thaler CD, Haimo LT (1996) Microtubules and microtubule motors: mechanisms of regulation. Int Rev Cytol 164:269–327

Trifonov EN, Berezovsky IN (2003) Evolutionary aspects of protein structure and folding. Curr Opin Struct Biol 13(1):110–114

Van Kampen NG (2007) Stochastic processes in physics and chemistry, 3d edn. Elsevier, Netherlands

Varma A, Palsson BO (1994) Metabolic flux balancing: basic concepts, scientific and practical use. Biotechnology 12:994–998

Whitton BA, Potts M (2002) The ecology of cyanobacteria: their diversity in time and space. Kluwer Academic Publishers, Dordrecht

Yeats CA, Orengo CA (2007) Evolution of protein domains. Encyclopedia of life sciences. Wiley, New York

Zeuthen T (1995) Molecular mechanisms for passive and active transport of water. Int Rev Cytol 160:99–161

Zhang L, Abbott JJ, Dong L, Kratochvil BE, Bell D, Nelson BJ (2009) Artificial bacterial flagella: fabrication and magnetic control. Appl Phys Lett 94:064107

Zwanzig R, Szabo A, Bagchi B (1992) Levinthal's paradox. Proc Natl Acad Sci 89:20–22. doi:10.1073/pnas.89.1.20

Index

A
Archaea, 17, 36, 85, 86–90

B
Biological cybernetics, 1, 4, 10, 26, 32

C
Cardiac cell, 36–39, 44, 48, 152
Controllability, 7, 8, 148, 150, 178, 193, 194

D
Directed movement of protocell, 4, 176, 181

E
Effectiveness, 6, 7, 9, 10, 24, 25, 29, 79, 117, 132, 142, 162
Effectiveness of energy conversion, 132
Energy transformation, 12, 136, 176, 182
Erythrocyte, 35, 59, 66

H
Hepatocyte, 35, 67–71
Homeostasis, 6, 8, 26, 35, 66, 115

I
Ion transport, 3, 17, 19, 20, 23, 25–27, 31, 35, 41, 43, 44, 47, 53, 56, 59, 63, 67, 70, 71, 78, 85, 91, 96, 103, 112, 118

M
Minimal movable cell, 182
Mitochondria, 2, 35, 36, 63, 71, 72, 74–77, 85, 127

N
Neuron, 46–47, 50–58, 68, 79, 164, 186

R
Regulation, 43, 56, 63, 70, 71, 112, 156
Robustness, 8, 9–11, 29, 31, 45, 63, 75, 142, 163, 187

S
Synthesis of transport subsystem, 13, 137, 138, 140, 148
Systems biology, 1, 2, 8, 11, 12, 59, 103

T
Thylakoids, 125–127, 176
Transport of large molecules, 131, 182, 183, 185, 186, 187
Transport of substances in biomembranes, 1, 26, 29, 78

V
Vacuole, 118, 122

Index

MIX
Papier aus verantwortungsvollen Quellen
Paper from responsible sources
FSC® C105338

If you have any concerns about our products,
you can contact us on
ProductSafety@springernature.com

In case Publisher is established outside the EU,
the EU authorized representative is:
**Springer Nature Customer Service Center GmbH
Europaplatz 3, 69115 Heidelberg, Germany**

Printed by Libri Plureos GmbH
in Hamburg, Germany